# GROUNDWATER TREATMENT TECHNOLOGY

# GROUNDWATER TREATMENT TECHNOLOGY

**Evan K. Nyer**

CBS

**CBS PUBLISHERS & DISTRIBUTORS PVT. LTD.**

New Delhi • Bangalore • Pune • Cochin • Chennai (India)

ISBN : 978-81-239-1900-3

First Indian Edition : 1987
Reprint : 2000



Reprinted in India by CBS Publishers & Distributors in arrangement with Van Nostrand Reinhold Company Inc., USA



Published by Satish Kumar Jain and produced by V.K. Jain for
CBS Publishers & Distributors Pvt. Ltd.,
CBS Plaza, 4819/XI Prahlad Street, 24 Ansari Road, Daryaganj,
New Delhi - 110002, India. • Website: www.cbspd.com
e-mail: delhi@cbspd.com, cbspubs@airtelmail.in
Ph.: 23289259, 23266861, 23266867 • Fax: 011-23243014

*Branches:*

- ***Bengaluru:*** Seema House, 2975, 17th Cross, K.R. Road,
  Bansankari 2nd Stage, Bengaluru - 560070
  • Ph.: +91-80-26771678/79 • Fax: +91-80-26771680
  • E-mail: cbsbng@gmail.com, bangalore@cbspd.com
- ***Pune:*** Bhuruk Prestige, Sr. No. 52/12/2+1+3/2,
  Narhe, Haveli (Near Katraj-Dehu Road By-pass), Pune - 411041
  • Ph.: +91-20-64704058/59, 32342277 • E-mail: pune@cbspd.com
- ***Kochi:*** 36/14, Kalluvilakam, Lissie Hospital Road,
  Kochi - 682018, Kerala • Ph.: +91-484-4059061-65
  • Fax: +91-484-4059065 • E-mail: cochin@cbspd.com
- ***Chennai:*** 20, West Park Road, Shenoy Nagar, Chennai - 600030
  Ph.: +91-44-26260666, 26208620 • Fax: +91-44-42032115
  • E-mail: chennai@cbspd.com

*Printed at* :
Nazia Printers, Delhi

# Preface

The purpose of this book is to give a broad understanding of treatment technology for contaminated groundwater. The cleanup of groundwater is a unique problem. Although a variety of technology exists for the removal of compounds from water, the manner in which these techniques are applied as treatment systems will have to be altered for groundwater cleanups. This book will give the reader a general understanding of contaminated groundwater and specific knowledge in the application of existing technology to cleaning up groundwater.

During the last several years, there has been increasing interest in groundwater. Several cities have had their water supplies contaminated with chlorinated hydrocarbons. Cleanups of abandoned hazardous waste sites, under Superfund, have shown that the aquifers below these sites have also been contaminated. In agricultural areas, pesticides and nitrates have been found in the groundwater. Leaking underground storage tanks have contaminated land and aquifers throughout the country. These are the cases that have been found in many areas and have had publicity. Many aquifer contaminations are unique and go unnoticed for years.

Where these problems have been found, wells have been drilled, samples taken, and models produced to show the extent of the contamination in the aquifer. A good volume of literature has been written and is available on drilling wells, sampling, and modeling groundwater. The people responsible for the cleanup—engineers, project managers, state and federal regulators, etc.—can go to the literature and educate themselves about these areas of technology. This is not the case for treatment of groundwater.

The treatment systems that have been installed have been emergency systems or experimental systems. There have not been enough cleanups to establish good engineering practices in the area. In addition, many of the cleanups have been private, and the details have not been released to the public.

This book will try to summarize the present knowledge and experience in the cleanup of groundwater. More importantly, the book will give engineers an understanding of how groundwater treatment systems are different from existing technology and of the methods that engineers must use to design a groundwater treatment system based on their present design knowledge.

The book will also provide, for the people who have general responsibility for the cleanup, a general knowledge in the different aspects of a cleanup. The goal of a cleanup cannot simply be to remove all of the contamination. The long-term solution must be permanent and economical. A "quick fix" may have problems two or three years after the cleanup. The most expensive method of cleanup may not be the best cleanup strategy for removing all of the contaminants. It is important for the design engineers and the engineers who regulate the aquifer cleanups to understand what is possible and what is economical on a groundwater cleanup.

This book will not be the final word in groundwater treatment technology. It is only a starting point. Groundwater cleanups have just begun. There will be many advancements in the near future in this area. It is hoped that this book will focus some of the thinking in this area and hasten the progress.

Many people contributed to making this book as accurate and as informative as possible. Some added specific knowledge in their area of expertise, and others added general review. I would like to thank Kevin Sullivan for his assistance on air stripping and Mark Stenzel for his assistance on carbon adsorption. They both helped in the main part of the book and provided case histories for Chapter 6 as well. I would also like to thank Doug Gore, Paul Flathman, and Gregory Githens for providing case histories for Chapter 6.

All of these gentlemen work for companies that provide specific technology to the groundwater area. They are in the forefront of the application of their respected expertise. To provide accuracy and realism, I have used specific names of products provided by these gentlemen. I have also provided specific names when discussing other areas of technology. This does not mean that either Van Nostrand

Reinhold or I endorse these specific products. These companies had specific knowledge that I hoped would be helpful to the reader, and they were willing to share their knowledge with the public. There are other companies in the field that provide the same services. Any cleanup project should contact several sources for specific technology and equipment.

In the general review area, I first have to thank my wife Betsy. No one will read this book as often as she has. If the reader finds the book easy to read, he or she has Betsy to thank for that. I would also like to thank Les Grady and Richard Conway for their comments on the book. Both have been friends for many years and were very helpful on the final draft of the book. Finally, I would like to thank Marcia Rajkovich for the artwork.

EVAN K. NYER

# Contents

**Preface / v**

**1. DEFINING TREATMENT PARAMETERS / 1**
Flow / 2
Influent Concentration / 7
Discharge Requirements / 11
References / 15

**2. LIFE-CYCLE DESIGN / 16**
Time Effect on Parameters / 17
Capital Costs / 27
Operator Expenses / 30

**3. TREATMENT FOR ORGANIC CONTAMINANTS: PHYSICAL/ CHEMICAL METHODS / 35**
Pure-Compound Recovery / 38
Air Stripping / 47
Carbon Adsorption / 61
Miscellaneous Treatment Methods / 80
References / 82

**4. TREATMENT FOR ORGANIC CONTAMINANTS: BIOLOGICAL METHODS / 84**
Biological Treatment / 85
In-Situ Treatment / 100
References / 109

**5. TREATMENT METHODS FOR INORGANIC COMPOUNDS / 110**
Chemical Addition / 112
Removal of Suspended Solids / 117

Miscellaneous Methods / 129
References / 132

6. **FIELD APPLICATION OF DESIGN METHODS / 133**
Investigation of Subsurface Gasoline Contamination, *Doug Gore* / 133
Air Stripping for a Municipal Groundwater Supply, *Kevin Sullivan* / 146
Low-Concentration Organic Removal for a Drinking Water, *Mark H. Stenzel* / 151
Biological Treatment of a Groundwater Contaminated with Phenol / 161
In-Situ Biological Treatment of Isopropanol, Acetone, and Tetrahydrofuran in the Soil/Groundwater Environment, *Paul E. Flathman and Gregory D. Githens* / 173
References / 184

**Index / 187**

# GROUNDWATER TREATMENT TECHNOLOGY

# 1 Defining Treatment Parameters

During the past five decades, environmental businesses and others have developed highly successful treatment techniques for the removal of contaminants from water. These methods have been developed in order to protect our national rivers, lakes, and other water bodies. Proven treatment methods are available for compounds ranging from human domestic wastes to toxic organic waste and heavy-metal comtamination.

During the past several years, the public has become more aware, and are now taking action, to protect one further body of water, groundwater. About one-half of the U.S. population relies on groundwater as its primary source of water (1). The main weapons in our fight against groundwater contamination will be the same methods that have been developed for other water bodies. Almost all of the compounds that we are currently finding in groundwater have already been found in wastewater; and a method for removing them has been developed, installed, and proven. In fact, for most compounds, multiple methods have been successfully applied.

However, we must be wary of direct broad application of wastewater treatment methods to groundwater. Although the techniques will be the same, the method that we will use to engineer and design these treatment methods will be different. The cleanup of groundwater is very different from the cleanup of wastewater.

The most obvious difference is that with groundwater cleanups, the body of water is actually being cleaned. In wastewater cleanups, we control and treat the wastewater that is entering a body of water. The body of water, a river or lake, actually cleans itself once we stop putting pollutants into it. Groundwater is not able to clean itself at a rapid rate. So, in a groundwater cleanup, we must clean up the source of the pollutants and we must also clean up the aquifer itself.

Other differences exist, and they will become obvious as we define the parameters for the treatment system design.

## FLOW

A good place to start in determining the parameters that will define the treatment techniques and final system design is flow. In wastewater treatment, the flow is a relatively simple parameter to determine. For domestic systems, the engineer selects a design date in the future. Most domestic systems are designed for 20 years into the future. The engineer then uses the estimated number of people living in the town in 20 years and multiplies that by a standard factor for per capita use of water. To this figure, the engineer adds any special flows from industrial waste in the city and, depending on the age of the sewer system, a flow factor for rain inflow and storm sewers.

Industrial waste flow is determined by adding up the expected flows from each of the unit operations at the plant. Once again, depending on the age of the plant, a factor is added for storm water runoff. In both cases, a safety factor of about 20% is added to the final figure. For both cases, flow is not a design variable; it is a set parameter that must be determined by the engineer.

In groundwater cleanup, flow is a design variable. The engineer does not determine the flow by standard formulas, but instead must weigh the effect of flow on the total cost of the system and on the time that is needed for a final cleanup. Let us look at the different factors that determine the flow for a groundwater treatment system.

Figure 1-1 shows a typical spill situation that has entered an aquifer. The contaminant travels through the unsaturated zone (the area above the aquifer not saturated with water), and encounters the aquifer. Groundwater, for the most part, does not sit still. The movement is relatively slow, 5 ft/yr to 5 ft/day (2), and is determined by the differential head between two points and the resistance to flow caused by the makeup of the aquifer (dense soil will be more resistant to flow than loose aggregate). The potential water head can be determined by simply drilling a well and measuring the depth of the water in the well.

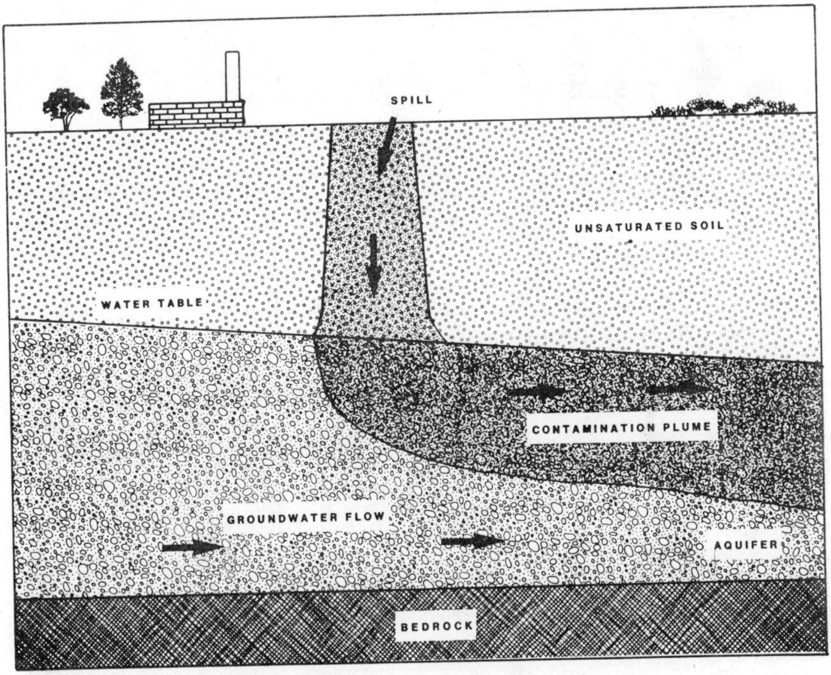


Figure 1-1. Contamination plume in an aquifer.

As can be seen in Figure 1-1, the contamination plume travels in the direction of the groundwater flow.

Flow for the treatment system is usually provided by sinking a well down to the aquifer and pumping contaminated water from it. As water is removed from a particular point in the aquifer, the water level is changed. Since flow in an aquifer is determined by the relative water levels, the flow for the treatment system will cause a change in the flow patterns in the aquifer. The area affected by this drawdown is called the "zone of influence" (Figure 1-2).

If the well is placed correctly and enough water can be drawn from the well, the movement of the contamination plume can be stopped and reversed. This will return the contamination to a point where it can be taken above ground to the treatment system. The first factor in the flow component of the treatment system design is the flow necessary to stop and reverse the movement of the contamination plume.

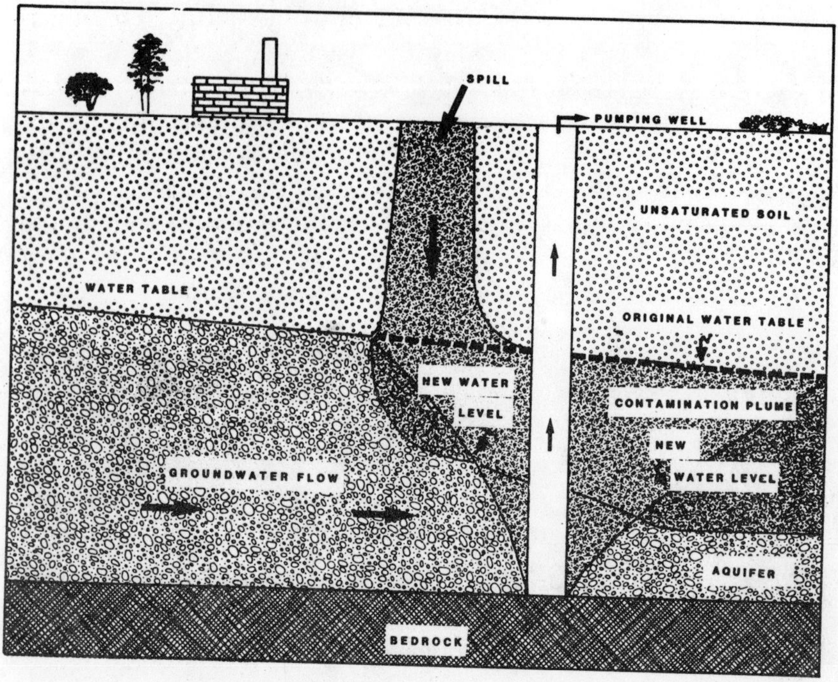


Figure 1-2. Zone of influence from a pumping well.

There are, however, other methods to stop the movement of the plume. An underground structure, clay slurry wall, or other device can be placed up-gradient of the contamination site or completely surrounding the contamination plume. When correctly designed and constructed, the slurry wall can prevent groundwater from entering the site. With no water movement through the contamination zone, the plume will not expand further for a relatively long time, perhaps 10–20 years.

The slurry wall is expensive, and the treatment design engineer will have to work with the hydrogeologists and other engineers and experts in the underground structures to determine the cost alternatives between stopping the plume movement physically or by water flow.

It should be noted here that the aquifers pictured in Figures 1–1 and 1–2 depict the "perfect" aquifer. In many cases, the bedrock on the bottom is cracked and is a source of water or drainage. Also, the aquifer can be so deep that the plume does not reach the bottom. It is beyond the scope of this book to discuss every possible type of underground situation. The basic idea is that the plume has to be stopped, and one factor in stopping the plume is water flow from the recovery well. All of the design factors discussed in this chapter will also be discussed in basic terms. The treatment design engineer will always have to work with the people who are expert on what is happening underground. The reverse is also true. The final cleanup design should not solely rest with the hydrogeologists. The two disciplines must work together to develop the most cost-effective alternative.

The second factor that may have to be considered as part of the flow is the amount of water entering the contamination area. One part of this has already been discussed. Slurry walls, or similar devices, can interrupt the flow into the contamination site from the aquifer. However, there are two other possible entry points. Surface water from rains and runoff can percolate down to the aquifer; or the bedrock below the aquifer can have cracks and fissures, providing a source of water.

Water from the surface can be controlled either by capping the contamination site with an impermeable layer or by providing good surface area drainage directed away from the site. Water coming from the bedrock is much harder to control and will probably have to be added to the flow for the treatment system.

The third factor in the flow to the treatment system is the speed of the cleanup. The more water that is drawn from the well, the lower the level of water surrounding the well. This will increase the head differential in the groundwater and force the water to the well at a faster rate. There is, however, a limit to how fast the water can travel through an aquifer. It is very easy to pump a well dry. Below this level, the engineer must decide between the cost of increasing the size of the treatment system and the savings by reducing the time for cleanup.

The relationship between flow and cleanup time is not perfect. For example, Figure 1-1 shows the contaminants flowing through the unsaturated zone to the aquifer. Let us assume that a well is placed in the contamination plume and that maximum flow over a period of three months is sufficient to completely eliminate the plume. The equipment is packed up and everyone leaves. The problem is that even if the original source of the contaminant has been controlled, there are still contaminants in the unsaturated zone. These can still migrate to the aquifer, and three months later, another contamination plume could form.

One solution to this problem would be to slow down the original cleanup so that the natural flushing methods have a chance to bring most of the contaminants through the unsaturated zone. Of course, an alternative would be to speed up the natural flushing action (see In-Situ Treatment, Chapter 4) and maintain the original speed of the cleanup.

Another example, one that cannot be solved as readily, is seasonal variation in the groundwater level. As one would expect, during times of high rain flow, spring and fall, the overall level in the aquifer can increase. The problem arises when the level recedes and the contaminants are left behind in the unsaturated zone. The next time the water level is high, the contaminants will reenter the aquifer. In these cases, the best solution may be to slow down the flow to the treatment system and perform the cleanup over several high-water seasons.

Other factors will have a relatively minor effect on the flow of the treatment system. The number of wells used will affect the amount of flow needed for cleanup. A well closer to the leading edge of the plume will require less water to stop the movement of the plume than a well at the site of the original contamination. However, it must be remembered that the idea is to clean up the groundwater. Therefore, a second well near the center of the contamination will also be required. The farther the plume has moved from the original contamination zone, the more likely will be the need for a second well or multiple wells.

Discharge requirements may affect the flow. More flow will be required if some of the water is to be used to flush the unsaturated zone of contaminants. Water returned to the site can decrease the time for cleanup by increasing the water head and forcing the groundwater to the central well at a faster rate. This same water will increase the total water entering the site, and the flow to the treatment system will increase accordingly. The effect of discharge requirements will be discussed further in the third section of this chapter.

In summary, the following factors have to be considered when the flow to the groundwater treatment system is determined:

1. Stopping/reversal of movement of the contamination plume

2. Amount of water entering the contamination site
3. The speed at which the cleanup is to occur
4. The number of wells to be used
5. The final disposal/use of the treated water

## INFLUENT CONCENTRATION

The concentration of contaminants in groundwater is normally determined by sampling the water from a well. The types of contaminants will depend on the material originally lost to the ground. The relative concentration will depend on where the well intersects the contamination plume. The farther away from the original release of contaminants, the more dilution will occur in the contamination plume. Several wells will have to be constructed to get a full picture of the plume. Once again, the treatment engir eer will have to work with the hydrogeologists to determine the influent concentration for the treatment system.

In addition to distance from the spill site, several other factors affect the concentration of the contaminants: amount of material reaching the aquifer, solubility and density, transformation of the compounds, amount of underground water flow, and time. These factors will all affect the size of the treatment system and the length of time that the system must be run for a full cleanup.

The factor concerning amount of material reaching the aquifer is made up of several components. The first is the amount of material lost to the ground. The first question to ask is whether the source of contamination has been shut off. There are certain cases in which the contamination is still being introduced into the ground. The primary example of this situation is a landfill.

This book will consider landfill leachate a source of contamination. Leachate in and of itself is not groundwater. The leachate must be collected and treated before it is allowed to come in contact with an aquifer. As with all sources of contamination, the leachate must be eliminated before the groundwater cleanup can proceed. It is counterproductive to start the cleanup while new material is still entering the aquifer.

Once the source of contamination is shut off, the engineer must determine the total amount of material lost to the ground. Not all of the material lost to the ground will reach the aquifer. The unsaturated zone above the aquifer will adsorb a percentage of the contaminants. It is widely reported that only 50% of the gasoline in a spill from a storage tank normally reaches the aquifer. Of course, this amount depends on the type of soil in the unsaturated zone, the distance to the aquifer, and the total amount of material spilled. The material left in the unsaturated zone will be discussed further in the next section of this chapter and under In-Situ Treatment in Chapter 4.

The next component of concentration in the aquifer comes from the solubility and the density of the material spilled. A large portion of the materials released into the ground are not soluble in water. Commercial gasoline has a solubility of only 20–80 mg/liter (3). When the material reaches the aquifer, it does not mix with the aquifer. Figure 1-3 shows a gasoline spill. The main component of the gasoline does not enter the aquifer. It floats on top of the aquifer and spreads in all directions. The floating material, however, will flow in the same direc-

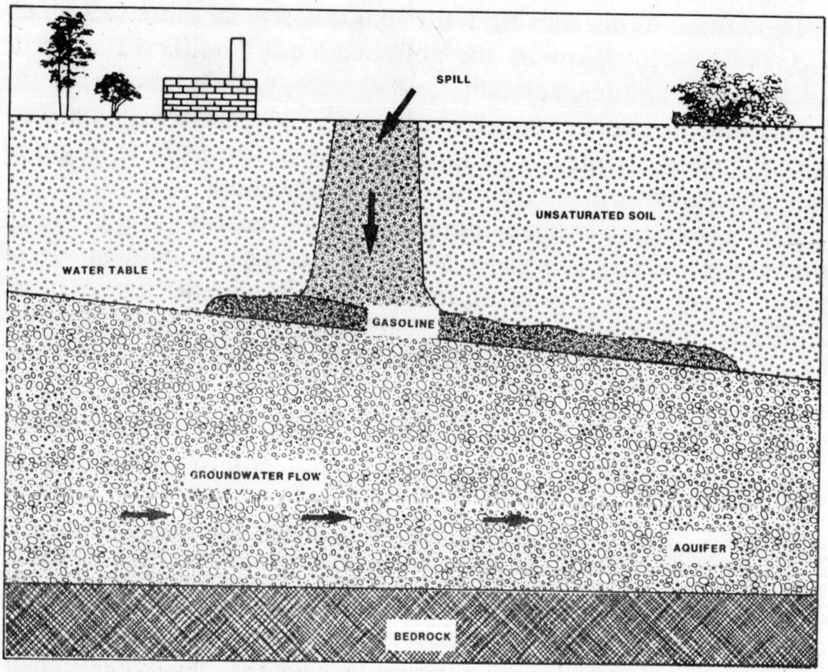


Figure 1-3. Gasoline spill encountering an aquifer.

tion as the groundwater. A small percentage of the gasoline, the ring compounds in unleaded gas and the alcohols in gasohol, does enter the aquifer and form the normal plume.

Gasoline is lighter than water, so it floats on top of the aquifer. Most straight-chain hydrocarbons are lighter than water and will stay on top of the aquifer. Compounds that are heavier than water will sink to the bottom of the aquifer. Figure 1-4 shows a trichloroethylene spill. Chlorinated hydrocarbons are generally heavier than water. Chapter 3 lists the specific gravity of several organic compounds. This list, combined with the solubility of the compound (solubility examples are also provided in Chapter 3), will tell the engineer where to find a particular compound once it is released into the ground. The section entitled Pure Compound Recovery in Chapter 3 will discuss finding and removing these compounds from the ground.

The contaminants can also be transformed in the ground. As stated before, the aquifer and the unsaturated zone are not able to clean

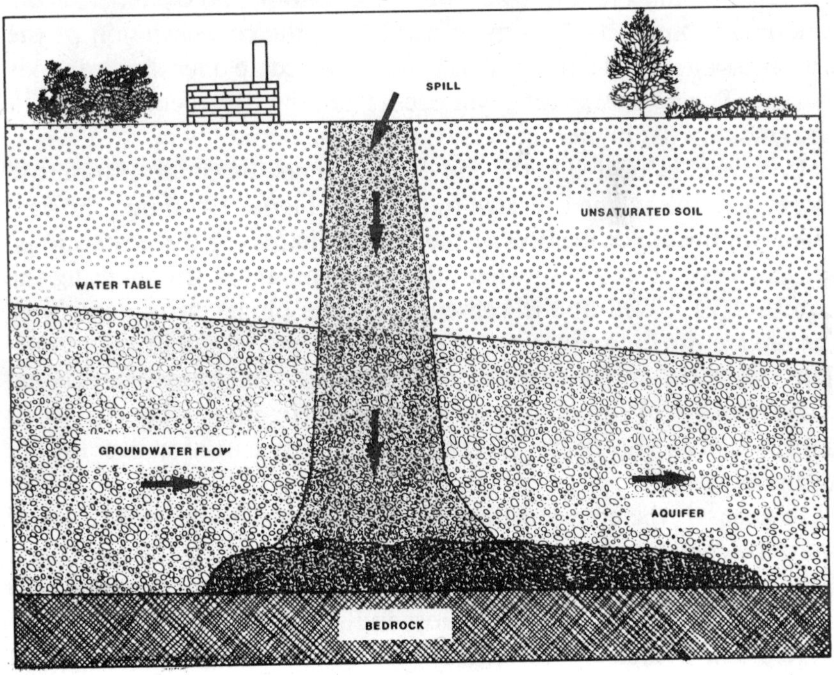


Figure 1-4. Trichloroethylene spill passing through an aquifer.

themselves in a reasonable period of time. However, there are biological and chemical reactions occurring below ground. Although these reactions are limited, they can affect the contaminants. The contaminants found in the aquifer and the unsaturated zone may be different from the contaminants originally released to the ground. This transformation can affect the nature and the structure of the compounds and, therefore, affect the concentration of the compounds in the aquifer.

Several minor factors combine to further affect the concentration of the contaminants. The faster the water flows past the point of entry of the contaminants, the lower the amount of contaminant per volume of water and the lower the concentration in the aquifer. There is also a tendency for the plume to spread and mix perpendicular to the direction of the flow. The mixing characteristics of the aquifer will affect the concentration.

Finally, all of these components combine with the treatment process to form a time effect on the concentration. The further away from the original spill site, the lower the concentration in the plume. As the plume is brought back to the central well, the concentration of the contaminants will decrease. This process is sped up once the cleanup is started. The normal mixing will occur, and the treatment system will be removing the highest concentration of material.

If the well is located at the end of the plume, the concentration will increase as the plume is drawn to the well. Figure 1–5 shows the time effect on the concentration of the contaminants for three situations: (a) leachate, or any case where the contaminant is still entering the aquifer; (b) a well at the beginning of the plume, with control of the source of contamination; and (c) a well at the end of the plume. The treatment design engineer must not only determine the original concentration of contamination from the different components listed here, but must also design for the change in concentration over the life of the project.

In general, the concentration found in groundwater will be lower than the concentration found in industrial or municipal wastewater. This is especially true for treatment of water found at the end of the contaminant plume. Whereas most industrial and municipal wastewaters are treated by biological methods, the low concentrations

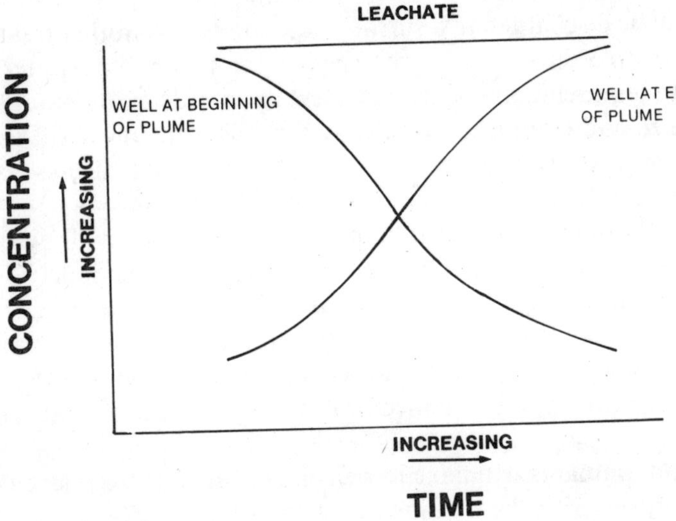


Figure 1-5. Time effect on concentration found in a well.

found in groundwater will make stripping and carbon adsorption the most widely used methods of treatment for groundwater.

In summary, the following factors have to be considered when the concentration to the groundwater treatment system is determined:

1. Distance of the sample well from the original entrance of the contaminant to the aquifer
2. The amount of material reaching the aquifer
3. The solubility of the contaminants in water
4. The relative density of the contaminants compared to water
5. Transformation of the contaminants
6. Rate of groundwater flow
7. Mixing characteristics in the aquifer
8. Combined time effect on concentration

## DISCHARGE REQUIREMENTS

The treatment plant design will also depend on the final disposition of the water that has been treated. The engineer must decide whether the water will be discharged to a surface water body, to another treatment system, or to a direct use, or returned to the ground or aquifer. The discharge requirements for each of these cases will have a major effect on the size and complexity of the system. The engineer will usually choose the least restrictive discharge, but may not always have a choice.

The preferred method of final disposal for the water drawn from the well is probably another treatment system, either a publicly owned treatment works (POTW) or an industrial treatment system. One of the advantages of the POTW is that the groundwater can be placed in any of the sewer lines. The advantage of the industrial system, if it is close, is that the type and concentration of the contaminants can be higher.

The contaminants will have to be compatible with the waste that is already processed at the treatment system. Some pretreatment may be necessary for either type of plant. The industrial plant may require neutralization, heavy-metal removal, or pure-compound removal. The POTW will usually require that the water entering the sewer not exceed the normal concentration of domestic waste. The following would be typical for a POTW: 250 mg/liter biochemical oxygen demand (BOD); 250 mg/liter suspended solids (SS); 100 mg/liter fats, oils, and grease (FOG); less than 1 mg/liter for any heavy metal; and between 0.5 and 5 mg/liter for specific, toxic organics. Regulations for each treatment plant will be different, and the potential plant will have to be contacted and the discharge limitations established.

The treated discharge may be sent directly to a stream or other surface water body. The following would be typical effluent requirements for a direct discharge: 30 mg/liter BOD, 30 mg/liter SS, less than 1 mg/liter for any heavy metal, and between 0.05 and 0.5 mg/liter for specific, toxic organics. A permit from the state environmental organization would probably be required for any direct discharge. The treatment plant would be considered a point source. The required monitoring and reporting with a direct discharge would add significant cost to the treatment system.

For an increasing number of situations, the spill has affected a drinking water source or other final use source. These cases require the most stringent treatment requirements. However, the contaminant concentration is normally very low or the well would be abandoned. Local drinking water standards should be referred to in order to determine discharge requirements. The U.S. EPA is scheduled to produce national drinking water standards sometime in the near future (4).

Finally, the water from the well can be returned to the ground. If the water is to be returned to the ground, it should be strategically placed to affect the movement of the plume. Figure 1-6 shows the water being returned at the end of the plume. This will increase the head and force the plume back to the central well.

The water can also be used to flush the unsaturated zone of contaminants. Figure 1-7 shows the treated water being returned at the surface where the spill originally occurred. This process, used to its maximum capabilities, is referred to as in-situ treatment. In-situ treatment will be discussed in Chapter 4. If the water is to be used at the end of the plume, the discharge requirements will be strict. This is because some of the water will not return to the central well. Any water escap-

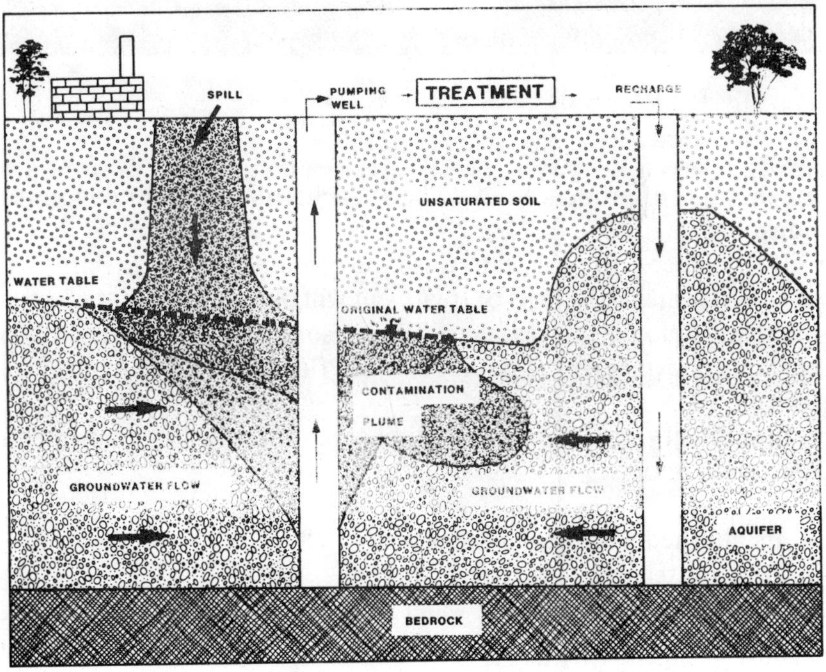


Figure 1-6. Effect of a recharge well on groundwater flow patterns.

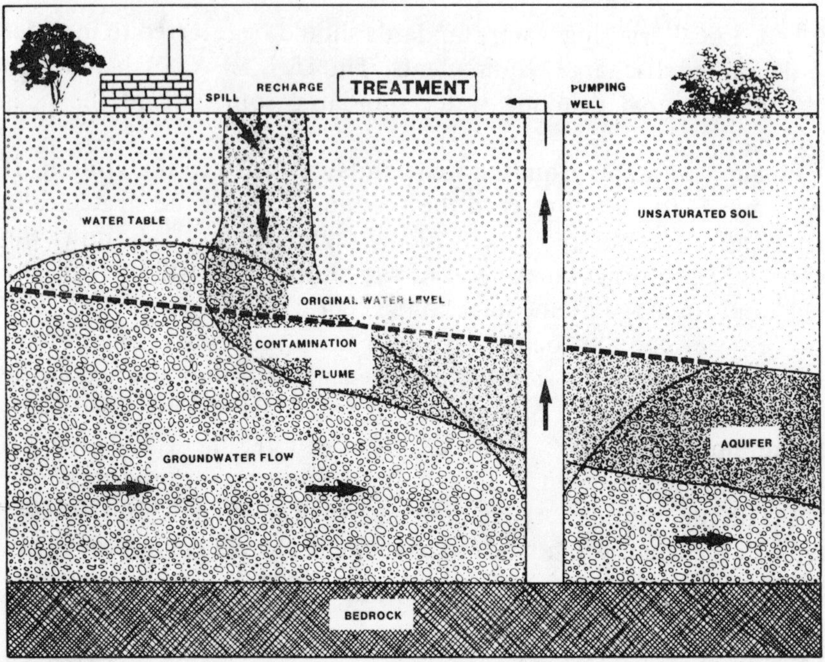


Figure 1-7. Effect of a surface recharge on groundwater flow patterns.

ing the zone of influence should be at background aquifer concentrations.

If the water is to be used to flush the contaminants or as part of an in-situ treatment, the level of treatment will be much lower. In fact, in the case of biological treatment, leaving some of the bacteria and the resultant enzymes will help solubilize and flush the contaminants from the soil.

In summary, the engineer can discharge the treated groundwater to four main outlets:

1. Established treatment system
2. Direct discharge to a river or stream
3. Direct use, as in drinking water
4. Returned to the aquifer via recharge wells or application to the soil

In addition, the water can be treated in situ. The pure application of this technology does not require the discharge of any water. However, in reality, in-situ treatment is usually a combination of in-place treatment and some discharge of treated water. Chapter 4 will cover this technology in detail.

After establishing the parameters or the choice of parameters, the engineer is ready to begin to design the treatment system. However, before going too far with the design, the engineer must take into account some of the unique properties that make up a groundwater treatment system design. Chapter 2 shows how such a system differs from a treatment system on an aboveground source.

## REFERENCES

1. Carter, K. B. Groundwater Contamination: A Pound of Prevention. *Journal Water Pollution Control Federation,* 56:7–810 (1984).
2. Anonymous. "Groundwater." Washington, D.C.: Chemical Manufacturers Association.
3. Freeze, R. A., and Cherry, J. A. "Groundwater." Englewood Cliffs, N.J.: Prentice-Hall, 1979.
4. Westrick, J. J., et al. The Groundwater Supply Survey Summary of Volatile Organic Contaminant Occurrence Data. Washington, D.C.: U.S. EPA, January 1983.

# 2
# Life-Cycle Design

As discussed in Chapter 1, most of the technology used for treatment of groundwater was developed for the wastewater market. The differences between gathering the design parameters for the two types of systems were discussed in that chapter. This chapter will discuss the changes on the equipment design itself caused by the unique situations we find in groundwater treatment systems.

Any wastewater treatment equipment is going to be constructed from materials that will last a minimum of 20–30 years. With regular maintenance on the movable parts (pumps, mixers, blowers, etc.), the process will produce good-quality water over that entire period. When engineers design a wastewater treatment plant, they design a system that will be viable for the next 20 years. The more water that is processed through the equipment (more years the equipment is in operation), the less it costs to treat each gallon of wastewater. The design is based on the life expectancy of the equipment, not the life expectancy of the project. This is why the design engineer looks 20 years into the future to determine the parameters for the wastewater treatment system. With the exception of treatment of drinking water, groundwater treatment systems will last much shorter periods of time, and the controlling parameter will be the life expectancy of the project.

In addition to the short life of the project, the systems will be relatively small. They will be so small, in fact, that the design engineer cannot merely use the designs developed for the wastewater market and simply shrink them. Other differences exist, and we will refer to all of these by describing the design method for a groundwater treatment system as *life-cycle design*. The most important factor to take into account in the correct life-cycle design is the time effect on the influent parameters.

## TIME EFFECT ON PARAMETERS

### Flow

The design engineer summarizes all of the factors for flow discussed in Chapter 1 and then establishes the design flow rate. This flow rate can be divided into two components: water sent off site and water recycled to the site. The amount of water discharged from the treatment site will probably remain the same during the life of the project. The amount of water that is being reused on site may vary during the cleanup. This variation will affect the flow through the treatment system. Water reused on site would include the water being used to increase hydraulic head at the end of the plume, to flush contaminants from the unsaturated zone, and for in-situ treatment. For these three cases, the flow to the treatment system may change during the project.

The flow would stay the same until the end of the cleanup. Near the end of the project, there would be less reason to continue using the water on site. Water is used to increase the hydraulic head at the end of the plume when the end of the plume is relatively far away from the central well and when the engineer wants to decrease the time for cleanup. Near the end of the project, the plume will have been forced back to the original contamination site. There will no longer be any reason to use water injection to help move the plume, although the engineer may still want to use water injection to speed up the progress of the project at the end.

The same thing may happen when the water is being used to flush the contaminants from the unsaturated zone. Near the end of the project, the contaminants will have been flushed and the cleanup of the aquifer will remain. At this point, the engineer may decide to eliminate the reuse of the groundwater. This is also the case for in-situ treatment. At the end of the project, the concentration in the aquifer may be low enough to stop recirculation and only remove the last of the contaminants. In all of these cases, the treatment plant engineer must take into consideration that the flow to the treatment system may change. The design engineer must make sure that the treatment system will function at all of the possible flow rates. The effect on the treatment system design will be discussed in the next section.

### Concentration

Chapter 1 discussed the time effect on concentration. Removal of the contaminants by the treatment system combines with the dilution effect, by clean water entering the affected site, to produce a steady decline in the concentration of contaminants in the groundwater when a well at the center of the plume is used for cleanup of the aquifer. A well at the end of the plume will have a steady increase in concentration as the plume is drawn toward the well. Figure 1–5 showed the changes in concentration over time for the two conditions.

The design engineer cannot assume that the lower the concentration is, the better the treatment system will operate. Certain processes are designed based on a minimum concentration. These units will lose efficiency with lower concentrations and finally stop functioning once a minimum concentration is reached. The design engineer must also take into account that the operational costs may be reduced as the concentration decreases. When the well is at the end of the plume, there is no reason to install equipment in the beginning that will not be needed for several years. When the life-cycle concentration increases, the treatment system may be installed over an extended period of time. In either case, the design must be able to accommodate the entire life-cycle concentration of the project. Three examples of the effect of concentration on unit operations follow.

Treatment methods for organic contaminants will be discussed in Chapters 3 and 4. Treatment methods for inorganic contaminants will be discussed in Chapter 5. The reader should refer to those two chapters for details on the following examples. First, let us look at the effect of time on an organic contaminant that we will treat with biological methods, specifically the activated-sludge method.

Assume that Figure 2–1 represents the influent life-cycle concentration as the groundwater is removed from a central well. The flow ($Q$) will be 25,000 gal/day for the entire life of the project. Also assume that all other environmental parameters are acceptable for biological treatment. Figure 2–2 pictures the proposed treatment system.

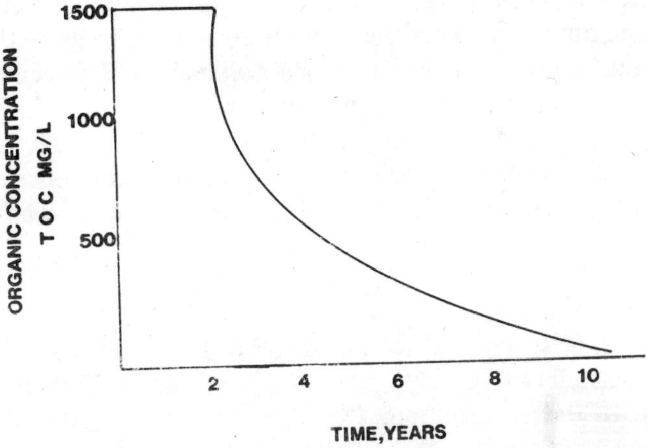


Figure 2-1. Life-cycle concentration from a well at the center of the plume for an organic contaminant.

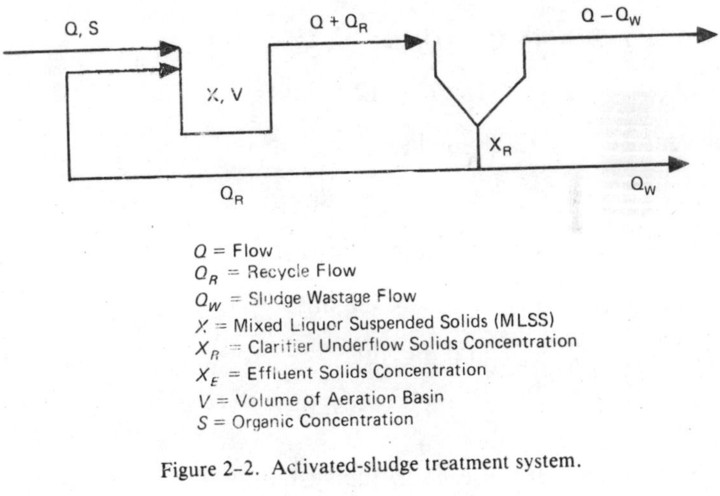


Figure 2-2. Activated-sludge treatment system.

As will be discussed in Chapter 4, the activated-sludge process depends on the settling properties of the bacteria that are growing. This process is based upon separating the residence time of the bacteria from the liquid residence time in the reaction (aeration) vessel. The bacteria settle in a quiescent tank (clarifier) and are returned to the aeration tank, with the treated water continuing on.

However, the specific gravity of the bacteria is very close to that of water. The only way to get the bacteria to settle is to grow them in a growth phase in which they produce polymers and flocculate. The combined mass then settles in the clarifier.

To keep the bacteria (MLSS) in a growth phase in which they settle properly, the bacteria should have a bacterial residence time or "sludge age" ($A$) between 5 and 20 days. Let us look at what happens to the sludge age during the life of the project.

Assume:

| | | |
|---|---|---|
| mixed liquor suspended solids, MLSS ($X$) | = | 3000 mg/liter |
| yield coefficient ($Y$) | = | 0.25 lb/lb |
| volume of the aeration tank ($V$) | = | 40,000 gal |

$$A = \frac{X \bullet V}{Q \bullet S \bullet Y} \tag{2-1}$$

For year 1, $S$ = 1500 mg/liter

$$A = 12.8 \text{ days}$$

For year 3, $S$ = 1200 mg/liter

$$A = 16 \text{ days}$$

For year 5, $S$ = 600 mg/liter

$$A = 32 \text{ days}$$

And for year 7, $S$ = 300 mg/liter

$$A = 64 \text{ days}$$

As can be seen from these data, the system will maintain the proper sludge age for about four years. After this time, the sludge age will be too high and the bacteria will lose their settling properties, and the clarifier will not be able to separate the bacteria from the treated water. Once the clarifier fails, the system will not be able to maintain a high concentration of bacteria in the aeration basin. At this point, the system will no longer remove a high percentage of the incoming organic contaminants.

One solution to this problem would be to lower the MLSS concentration. Figure 2–3 summarizes the sludge ages for the treatment system at MLSS levels of 3000 mg/liter and 1500 mg/liter. Although this does extend the useful life of the treatment system, the system still fails before the cleanup can be completed. There is also a lower limit to the MLSS. The MLSS concentration entering the clarifier must be about 1250 mg/liter or above to insure proper settling. Bacteria rely on flocculation in order to settle, and critical mass is required to insure enough contacts between the flocculating particles.

Another method to extend the useful life of the system would be to divide the aeration basin into two or more tanks. In our example, we could use two 20,000-gal tanks instead of the one 40,000-gal tank. Assuming 1500 mg/liter MLSS, at year 6 one aeration basin could be

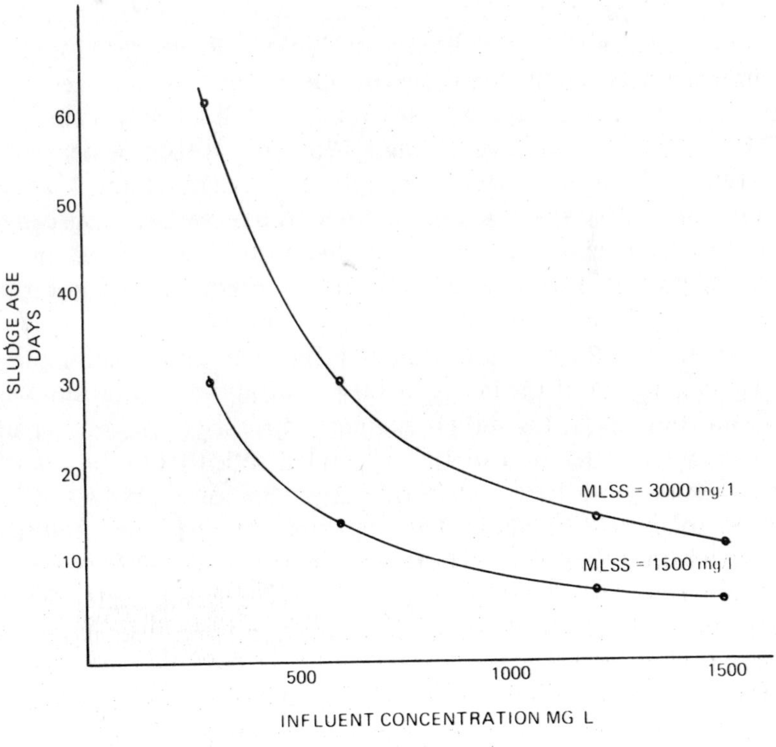


Figure 2–3. Sludge age with life-cycle influent concentrations.

shut down. This would effectively halve the sludge age in the system at a steady MLSS. An added advantage of this method would be that half of the blowers could also be shut down. The system would not only last longer, it would also cost less to run in the final years of operation.

Of course, there are limitations to an activated-sludge system designed this way. The final few years of the cleanup will still create a very long sludge age. Similar limitations will affect fixed-film, biological treatment system designs. The actual design may have to include different unit operations to clean up the groundwater over the entire life-cycle (see Biological Treatment of a Groundwater Contaminated with Phenol in Chapter 6 for an example of life-cycle design using biological methods in conjunction with other treatment methods). The change in concentration of contaminants over the life of the project may have a detrimental effect on the performance of the treatment system. The design engineer must take into account the entire range of concentrations when designing the treatment system.

These problems are not limited to biological systems. Inorganic treatment systems have life-cycle considerations. For our second example, let us take an aquifer contaminated with a heavy metal.

The design details for removing a heavy metal from water will be covered in Chapter 5. For this example, assume that if lime is added, to a pH of 9, all of the heavy metal comes out of solution. Suspended solids, metal hydroxides, are formed that must be separated from the water. Figure 2–4 summarizes the life-cycle concentration for the heavy metal. Assume that laboratory analysis is performed and Figure 2–5 summarizes the concentration of suspended solids resulting from lime precipitation of the heavy metal at various concentrations.

From these data, the design engineer must select the proper unit operation for separation of the suspended solids from the groundwater. During the first couple of years, the concentration of suspended solids will be about 4000 mg/liter. At this range, the main problem with separation will be the thickening of the solids as they are removed from the water. A standard clarifier (see Chapter 5 for a description of a clarifier), with its design based on thickening, is the proper unit operation.

After year 3, this is no longer the case. Metal hydroxide suspended solids rely on flocculation to be removed from the water. As in ac-

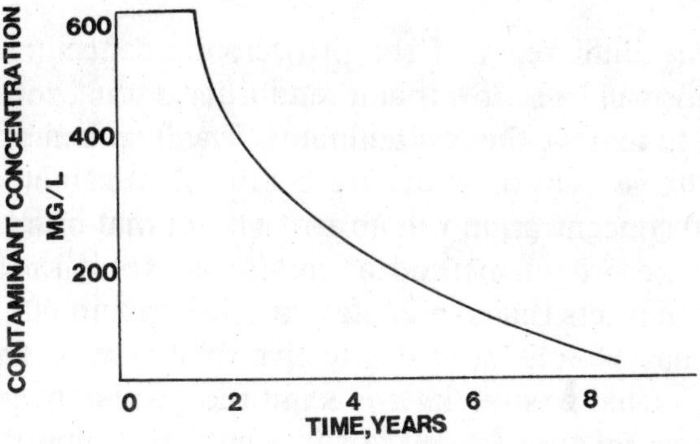


Figure 2-4. Life-cycle concentration from a well at the center of the plume for an inorganic contaminant.

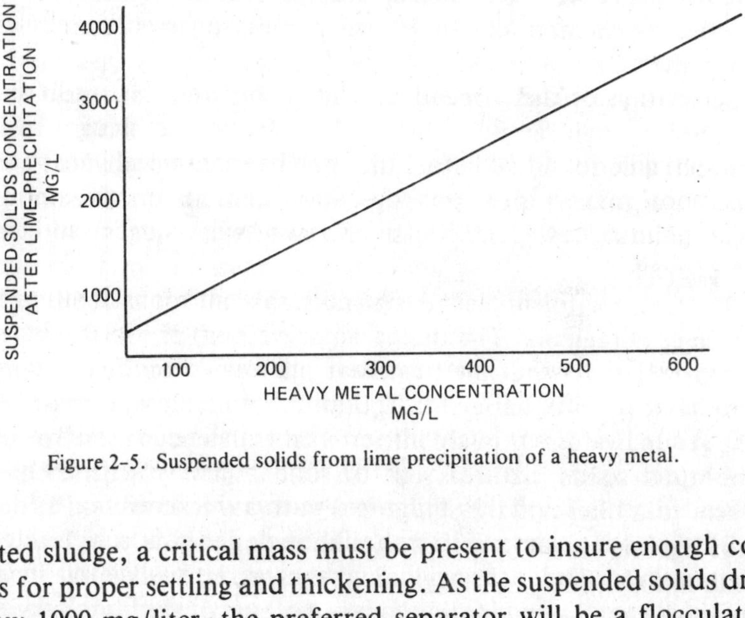


Figure 2-5. Suspended solids from lime precipitation of a heavy metal.

tivated sludge, a critical mass must be present to insure enough contacts for proper settling and thickening. As the suspended solids drop below 1000 mg/liter, the preferred separator will be a flocculating clarifier (see Chapter 5). As the solids progress to less than 500 mg/liter, a solids contact clarifier (see Chapter 5) will be required. The flocculating clarifier has a chamber that increases the number of contacts between floc particles before entering the clarifier zone. The solids contact clarifier contacts previously settled solids with incoming solids to obtain the critical mass for settling.

During the final years of the project, the concentration of suspended solids will be so low that a sand filter or dual-media filter will be required to remove the contaminants. The filter is the proper selection after the suspended solids are below 100 mg/liter. Finally, the heavy-metal concentration will be at the point that lime precipitation will not be the proper method of treatment. Once the heavy-metal concentration is less than 5 mg/liter, another technology, such as ion exchange, may be a more cost-effective treatment technique. Selection of ion exchange over lime precipitation will depend upon total flow to the system and effluent requirements. (It should be noted here that all of the concentration values given in this example are approximations. Every heavy metal and every groundwater will react differently with lime precipitation. Only through laboratory testing can the design engineer determine when these different levels will be reached.)

Separation of the suspended solids is not the only problem that the design engineer must face. Metal hydroxide sludges must be dewatered or solidified before they can be economically disposed of. Once again, the proper unit operation depends on the amount of solids to be processed. Of course, this value will change over the life of the project.

The final treatment plant design must take into consideration all the above requirements. The design engineer must try to maximize the number of years that the treatment plant will function. Figure 2-6 shows one possible design incorporating the life-cycle considerations. The groundwater is brought into contact with lime in a flash mix tank. The liquid/solids mixture is sent to a solids contact clarifier. The water is sent to a filter and the sludge is sent to a thickener.

The solids contact clarifier should be designed at a high rate. The high solids loading in the beginning years can be quickly transferred to the thickener. The solids that escape because of the high loading will be captured by the filter. During the middle years, the solids contact clarifier will maintain the critical mass necessary for proper settling.

The filter should be a dual-media design. This design can handle a higher concentration of suspended solids. In the beginning years, the filter will remove the solids from the overloaded solids contact clarifier. During the final years, the clarifier can be shut down and the filter will remove all of the suspended solids.

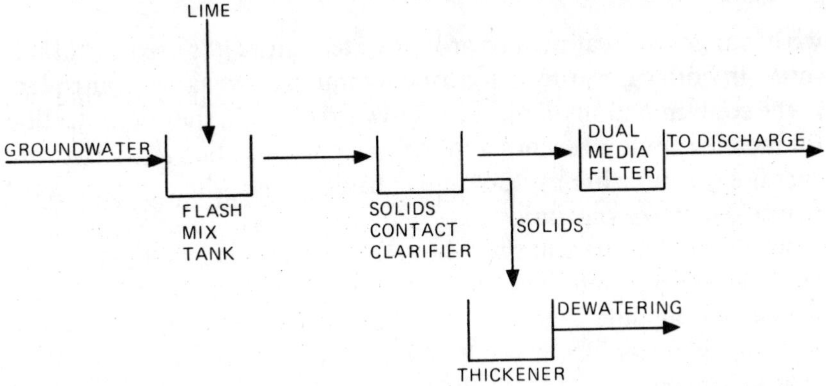


Figure 2-6. Life-cycle treatment plant design for a heavy metal.

The thickener will handle the high load of solids during the first several years. During the middle and end years, the thickener will act as a storage tank so that the solids dewatering activities can be run on a periodic basis. The solids dewatering will have to use extra manpower and equipment during the beginning years and will have to be scheduled during the middle and final years.

Finally, when the concentration reaches a very low level, all of the equipment in Figure 2-6 will have to be shut down. An ion exchange unit will be brought in and run until the last of the heavy metals have been removed from the groundwater.

Of course, Figure 2-6 represents only one possible design. Local conditions may favor a different solution to this design problem. It would be impossible to list all of the possible different designs for this problem. The point is that it is important for the design engineer to use the life-cycle considerations when designing the treatment plant.

There is also the possibility that the concentration will increase during the life cycle of the project. In the examples above, the well used for withdrawing water from the aquifer was at the center of the contamination plume, and the source of contamination was being controlled. This will be the case when the purpose of the well system is to clean the aquifer. However, in an increasing number of cases, the source of contamination is not known, and the purpose of the treatment system is to make the groundwater fit for consumption.

The papers on carbon adsorption and stripping in Chapter 6 give two examples of treatment of groundwater before it is used as potable water. In both cases, the source of contamination was not controlled. As the contamination plume moves toward the well, the concentration will increase until it reaches a maximum concentration. The engineer must use the same life-cycle design considerations when designing for an increase in concentration.

One final example will show how the concentration can affect the life-cycle design when the concentration is increasing. For this example, we will assume that the well, currently being used as a source of drinking water for a small town, has been contaminated with trichloroethylene.

The present concentration in the well is 5 parts per billion (ppb). Testing wells, upflow from the original well, show that the concentration of contaminant increases to a maximum of 50 ppb one mile from the original well. Modeling, of the aquifer and the amount of water being drawn for the town, shows that the original well will reach the maximum concentration in eight years. The drinking water regulations state that 5 ppb of this particular chlorinated hydrocarbon is safe for consumption, but the town has decided that they would like to get the concentration as low as economically feasible. There is no other aquifer available for the town's source of water.

The town hires a consultant to run a series of tests to determine the design data for carbon adsorption and air stripping. The carbon is able to remove the trichloroethylene to less than 1 ppb. The system design for an influent concentration of 5 ppb is a single column with a 30-minute residence time. The total costs, capital and operating, for this system is \$0.22/1000 gal. When the influent reaches 50 ppb, a second carbon column is required in order to get the maximum use of the carbon and maintain a low effluent concentration. The total costs for the high-influent concentration system is \$0.55/1000 gal.

The air-stripping system can approach the 1-ppb effluent level of trichloroethylene, but it has a difficult time consistently going below that level. Another problem with the air-stripping system is that the construction cannot be broken into two sections. Although the system required for treatment of the 50-ppb influent is much larger than the system necessary for the 5-ppb influent, the entire system must be installed at the beginning of the project. The total cost of the air-stripping system for the 50-ppb influent is \$0.16/1000 gal.

The town prefers the effluent concentration that the carbon can obtain. Also, the carbon can be on site within one week, whereas the air-stripping system will take three months to install. The initial differential in cost is also acceptable to the town, but the higher cost at the end of the project is considered too expensive.

The solution to this problem is that the town installs the carbon adsorption system. When the concentration reaches the point where a second carbon column is required, the town will install an air-stripping pretreatment system instead. The effluent from the air-stripping pretreatment unit only has to be 5 ppb. At that effluent concentration, the air-stripping unit will only cost \$0.07 / 1000 gal. The town gets the effluent concentration that they desired, and the life-cycle design keeps the cost at a reasonable range. Of course, the project may fall under Superfund and the federal government would pick up 90% of the initial capital cost. In that case, the entire treatment system would have to be purchased up front. Life-cycle considerations would still be valid. The town would be able to save the operational costs until the pretreatment unit was needed.

Life-cycle design problems will exist on other unit operations, although, for certain treatment systems, the effect will be more on operational costs than on physical design and possible failure. The design engineer must consider the operational cost over the entire project when developing the design.

## CAPITAL COSTS

As we discussed in Chapter 1, the total time for a cleanup would usually be less than the 20 years set for the life of a wastewater project. In the previous section of this chapter, we saw that even if the life of the project is 10 years, it would probably be unnecessary to use all of the equipment for the entire time. In this section, we will discuss the effect of time on the cost of equipment.

Most of the equipment used in the field will have a 5- to 20-year life expectancy. Pumps and other equipment with moving parts will have a lower life expectancy, and tanks and reaction vessels will have a longer life expectancy. The cost of equipment in wastewater treatment is figured over the life expectancy of the equipment. The cost of equipment on a groundwater cleanup must be based on the time the equipment is used on the project.

Assume that the cost of equipment for the activated-sludge treatment system example in the previous section was $100,000. If we set the amount of time that we need the equipment and the interest rate that we have to pay for the equipment, we can then calculate the daily cost of the equipment. One formula for calculating costs would be

$$C = \frac{Cap}{[1-(1+i)^{-n}]/i} \tag{2-2}$$

where:

| | | |
|---|---|---|
| $C$ | = | cost per time period $n$ |
| $Cap$ | = | capital cost ($100,000 in our example) |
| $i$ | = | interest rate per time period |
| $n$ | = | time |

We will assume that the interest rate is 12%. If the equipment is used for 10 years, the daily cost is $48/day. If the equipment is needed for only 5 years, the daily cost is $74/day. At 2 years, the daily cost is $157/day, and at 1 year, the daily cost is $296/day. Figure 2–7 summarizes the daily cost of equipment when used for various periods of time.

As can be seen, the cost of equipment gets significantly higher as the time of use decreases. The normal method of comparing the cost of treatment by different technologies is to base the comparison on cost of treatment per gallon of water treated. At a flow of 25,000 gal/day, the cost of treatment goes from $0.00192/gal at 10 years to $0.01184/gal at 1 year. Using the same equipment for 1 year will cost six times as much per gallon treated as using the same equipment 10 years.

A great many groundwater cleanups will be completed in a one- to two-year period. This makes the cost of equipment over time another part of the life-cycle design. The design engineer will have to consider life-cycle effects on the shorter projects and on the longer projects in which a particular piece of equipment is needed for only a short period

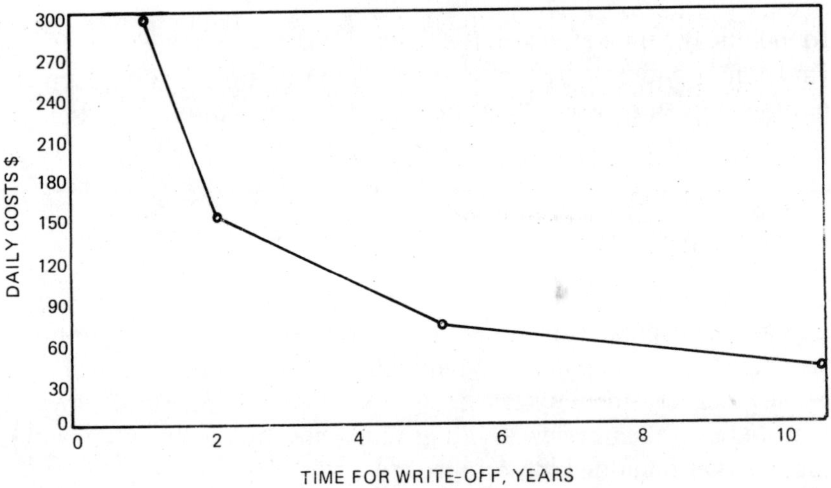


Figure 2-7. Capital cost as a function of time for write-off of equipment.

of time. Two obvious solutions to short-term use would be either to rent the equipment or to use the same equipment on several different projects. This would allow the equipment to be capitalized over 10 years, even though it was only required for a short period on each particular project.

Of course, any equipment that is to be used over more than one project will have to be transported from one project to the next. The equipment will have to be portable. For example, the design engineer has a choice of one tank 17 ft in diameter or two tanks 12 ft in diameter. If the equipment is to be used for only a short period of time, the proper choice is the two 12-ft-diameter tanks. The legal limit for a wide load on a truck is 12 ft. Although one 17-ft tank would be less expensive than two 12-ft tanks, the two 12-ft tanks could be shipped to the next project when the present project was completed. The two 12-ft tanks could be written off over a longer period of time. Therefore, the daily costs of the two tanks would be less than the one 17-ft tank. Materials of construction, internal moving parts, etc., will also have to be considered in the design of portable treatment equipment.

There are other ways to overcome the life-cycle problems of capital equipment. Once again, the important point is for the design engineer to include the time effect on the design. If the design engineer cannot find rental equipment, and there is only one project on which the equipment can be used, then the engineer's calculations should use a residual value for the equipment at the end of the project.

## OPERATOR EXPENSES

One final area that has to be discussed under life-cycle design is operator expenses. Any system that requires operator attention will cost more to operate than a system that does not require operators. All wastewater treatment system designs should have operator expenses factored into their review. With groundwater treatment systems, this factor takes on added importance.

The main reason for this importance is the relative size of a groundwater treatment system. Once again, the engineer cannot simply take a design developed for wastewater treatment systems and reduce its size for groundwater treatment. Most groundwater treatment systems will be small in comparison to most wastewater treatment systems. The operator costs, therefore, become more significant.

The biological treatment system example will be used once again. Assume that a 15-hp blower is required for the system at $0.06/kwh. In addition, chemicals and miscellaneous costs are $3.00/day. At a 10-year life for the equipment, the daily costs would be as follows:

| | |
|---|---|
| Equipment | $48.00 |
| Power | $29.00 |
| Chemicals | $ 3.00 |
| Total | $80.00 |

Figure 2–8 summarizes the relative costs for each category. Without any operator attention, the equipment represents 60% of the daily cost of operation. The power is 36%, and the chemicals are 4% of the daily costs. Figure 2–9 shows what happens to this relationship if one operator is required for one eight-hour shift per day and is paid, including benefits, $10.00/hr. Now 50% of the daily cost is represented by operator costs. Equipment drops down to 30%, power to 18%,

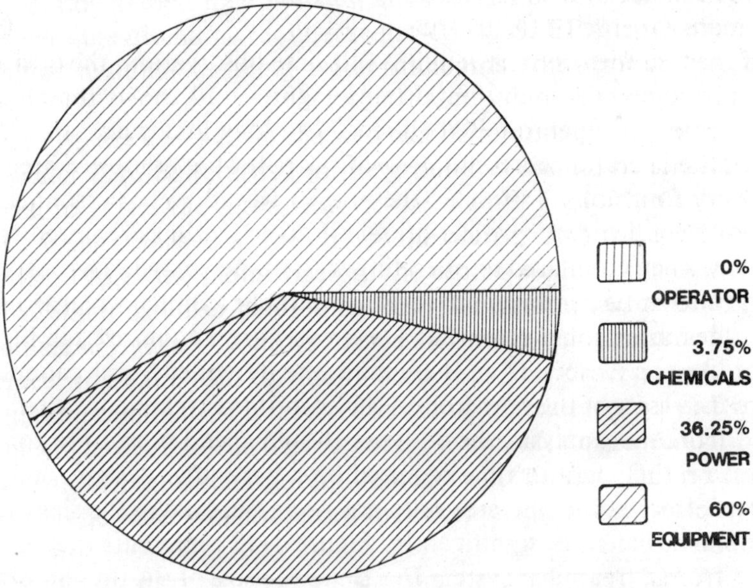


Figure 2-8. Ratio of daily costs with no operator attention.

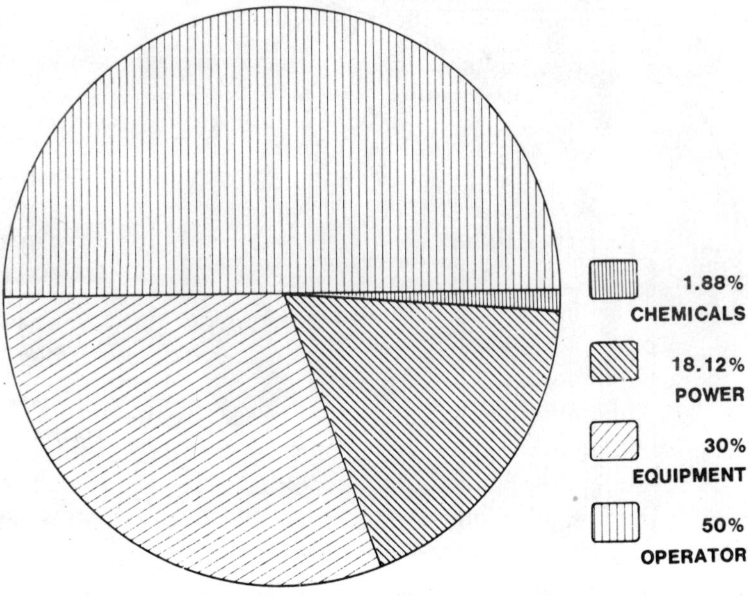


Figure 2-9. Ratio of daily costs with 8 hr/day of operator attention.

and chemicals to 2%. At only one shift per day, the operator is now the main expense of the treatment system.

If the treatment system requires full-time observation, the operator costs become even more important. Figure 2–10 shows the relative costs when an operator is required 24 hr/day and paid $10.00/hr. Now, the operator represents 75% of the cost of operation. Three out of every four dollars spent on the project would go to personnel.

Daily costs for the project double if an operator is required for 8 hr/day when compared to operating with no personnel. The costs triple at two shifts per day, and it costs four times as much when around-the-clock attention is required. These costs are summarized in Figure 2–11. As can be seen from these data, the design engineer cannot ignore the effect of the operator on the cost of the treatment system. In fact, from this analysis, the designer should spend most of his or her effort on the operator time required on a particular design.

The effect of the operator does not decrease, even as the size of the equipment increases significantly. Figure 2–12 represents the relative costs from a treatment system five times the size of the present exam-

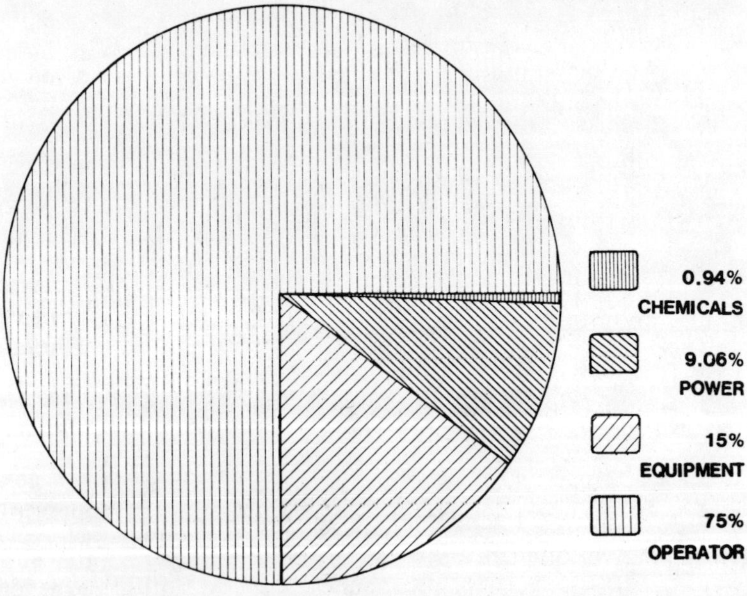


Figure 2–10. Ratio of daily costs with 24 hr/day of operator attention.

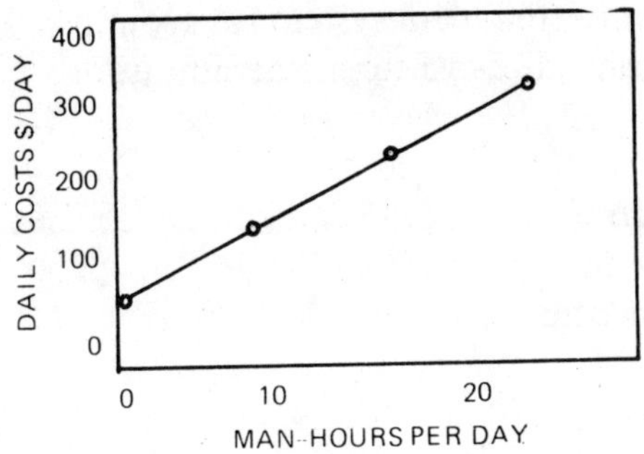


Figure 2-11. Daily costs of treatment with variable operator attention.

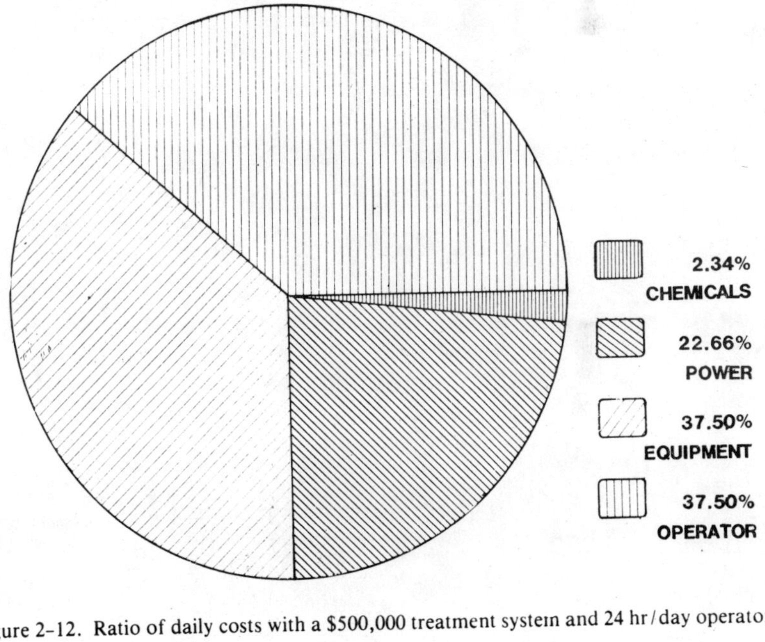


Figure 2-12. Ratio of daily costs with a $500,000 treatment system and 24 hr/day operator attention.

ple and requiring 24 hr/day of operator attention. The operator still represents over one-third the cost of treatment. Even as the total cost of the treatment system approaches $500,000, the design engineer must take special precautions to keep the required operator attention to a minimum.

In summary, the design engineer must use the life-cycle design methods to design a treatment system for groundwater. The following major areas all have life-cycle considerations:

1. Flow
2. Concentration
3. Capital costs
4. Operator expenses

# 3
# Treatment for Organic Contaminants: Physical/Chemical Methods

When a pure organic contaminant is released onto or into the ground, the main force on the movement of the compound is gravity. If the ground is porous, the spill will move downward. There will be some lateral spread of the movement, controlled by the porosity of the soil. The speed of movement will be dependent on the viscosity of the material spilled and the porosity of the soil. Three things can happen to the downward progress of the contaminant:

1. Adsorption on the soil particles
2. Stoppage by an impermeable layer of soil (clay, bedrock, etc.)
3. Encounter with an aquifer

As the contaminant moves in the soil, each soil particle will adsorb a small amount of the material. The material adsorbed stays with the soil and no longer moves with the main flow of the contaminant. If the spill does not reach an impermeable layer or aquifer first, the spill will eventually exhaust itself in the soil and stop all movement.

The amount of soil required to adsorb all of the spilled material depends upon two factors: the porosity of the soil, and the adsorbability of the contaminant as reflected by its characteristic *maximum residual saturation*. When the contaminant is at or below its maximum residual saturation, it will not move in the soil. The American Petroleum Institute (1) recommends the following equation to relate the amount of soil required to immobilize the contaminant:

$$C = \frac{0.20 \bullet V}{P \bullet Sr} \tag{3-1}$$

where:

- $C$ = cubic yards of soil required to immobilize the contaminant
- $V$ = volume of contaminant in barrels
- $P$ = porosity of the soil
- $Sr$ = residual saturation

Although this equation was originally developed for petroleum-based compounds, the relationships expressed by the formula are valid for other compounds. The material that is adsorbed in the soil has a greatly reduced chance of contaminating the groundwater. The portion of the material that is soluble in water will be picked up by rainwater on its movement through the ground and will still be able to contact and contaminate an aquifer. The rest will remain above the water.

Current thinking is that the adsorbed material does not represent a significant environmental risk, and the material is left in place. The main exception occurs when the spilled material is a volatile compound, as in gasoline spills. These compounds can move upward in the soil and enter structures through basements, producing explosion hazards or just nuisance odors (2). In these cases, something must be done about the spilled material in the unsaturated zone. Also, the future trend is to clean up all the material of a spill, not just the material that reaches the aquifer.

The problem lies in the fact that the material adsorbed on the soil particles is very difficult to find and remove. First, there is very little lateral movement in the unsaturated zone. The well must be constructed near the original place where the compound was released, if that site is known. Second, in addition to the components of the spill changing due to volatilization and solution in water, the compounds can also be transformed by the natural biota in the soil. Bacteria in the soil can change the structure of an organic compound. Therefore, the organics found in the soil may be different from the original organics released.

Once found, the material can be removed in several ways. For shallow spills, the ground can be excavated and either land farmed (if the organic is degradable) or buried in a controlled landfill. For deeper contaminations, flushing and recovery or in-situ treatment are the only methods of treatment. These will be covered in Chapter 4.

Many of the same problems are encountered when the spill's movements are stopped by an impermeable layer of soil. When the contaminant reaches the impermeable layer, its downward progress stops. The spilled material spreads out on top of the impermeable soil like a pancake. This process continues until either the contaminant contacts enough soil to adsorb the entire spill or the impermeable layer ends.

The same problems exist for a cleanup of material that hits an impermeable soil layer, as in the cleanup of an adsorbed material. It is hard to gather the material in one spot so that it can be recovered or treated. A slurry wall can be installed to prevent the continued spreading of the contaminant. Also, a well could be drilled down to the impermeable soil and the material recovered. The slurry wall would have to completely surround the spill, which would be very expensive. The well would not be able to recover a significant portion of the material spilled. Once again, some type of flushing or in-situ treatment are the best methods of cleanup.

Finally, the spill can reach an aquifer. Depending on the solubility and the density of the contaminant, the spilled material can do three things once it reaches the aquifer:

1. Float on top of the aquifer
2. Dissolve into the aquifer
3. Sink to the bottom of the aquifer

Pure-compound release is only one way in which contaminants can enter the aquifer. Leachate from a landfill or leaks from a storage or treatment pond are other sources of contaminants. Sometimes the original source of contaminants is never known.

In all of these cases, the aquifer must be cleaned. The treatment method or methods used for an organic cleanup will depend upon several factors. All of the following will have to be considered when the unit operations to be used are chosen:

A. Description of the spill
   1. Concentration
   2. Quantity of contaminant
   3. Total time allotted for cleanup
   4. Final use of the water
B. Properties of the spilled material
   1. Solubility
   2. Density
   3. Stripability
   4. Adsorbability
   5. Degradability

Other factors will be important on individual projects. This chapter will discuss the physical/chemical methods of removing organics from an aquifer. Chapter 4 will discuss biological methods and in-situ treatment. Chapter 5 will discuss inorganic treatment techniques. Finally, Chapter 6 will combine the design considerations presented in Chapters 1 and 2 with the treatment techniques presented in Chapters 3, 4, and 5 and provide examples of groundwater cleanups.

## PURE-COMPOUND RECOVERY

Pure-compound recovery is possible only when the contaminant is not soluble in water. Once the organic is in solution, the pure-compound recovery techniques will not work. This also includes the situations when a surfactant or similar compound creates an emulsion. Table 3–1 gives the solubility for 20 organic compounds. The specific gravity, Henry's law constant (stripability), carbon adsorption capacity, and degradability on these same 20 compounds will be provided in their respective treatment sections.

Straight-chain hydrocarbons are generally not soluble in water. Most petroleum products, oil, gasoline, etc., fall into this category. For soluble compounds such as acetone and phenol, pure-compound recovery techniques cannot be used.

**TABLE 3-1. Solubility of Specific Organic Compounds in Water.**

| COMPOUND | MILLIGRAMS COMPOUND/LITER WATER (@ °C TEMPERATURE) |
|---|---|
| 1. Acetone | Infinite |
| 2. Benzene | 1780 (20) |
| 3. Carbon tetrachloride | 800 (20), 1160 (25) |
| 4. Chloroform | 8000 (20), 9300 (25) |
| 5. Methylene chloride | 20,000 (20), 16,700 (25) |
| 6. Chloro benzene | 500 (20), 488 (30) |
| 7. Ethyl benzene | 140 (15), 152 (20) |
| 8. Hexachloro benzene | 0.11 (24) |
| 9. Ethylene chloride | 9200 (0), 8690 (20) |
| 10. 1, 1, 1-Trichloroethane | 4400 (20) |
| 11. 1, 1, 2-Trichloroethane | 4500 (20) |
| 12. Trichloroethylene | 1100 (25) |
| 13. Tetrachloroethylene | 150 (25) |
| 14. Phenol | 82,000 (15) |
| 15. 2-Chlorophenol | 28,500 (20) |
| 16. Pentachlorophenol | 5 (0), 14 (20) |
| 17. Toluene | 470 (16), 515 (20) |
| 18. Methyl ethyl ketone | 353 (10) |
| 19. Naphthalene | 32 (25) |
| 20. Vinyl chloride | 1.1 (25) |

[a]From Verschueren, Karel. "Handbook of Environmental Data on Organic Chemicals." New York, Van Nostrand Reinhold, 1983.
*Note:* Numbers in parentheses = temperatures.

If the compound is not soluble, then it will either float on top of the aquifer or sink to the bottom. The compound's relative position to the water in the aquifer will depend upon the relative density of the material. Organic compounds that are lighter than water will float, and organic compounds that are heavier than water will sink.

Table 3-2 provides the specific gravity of several organic compounds. Water is used as the reference compound for the specific gravities. Therefore, any compound with a specific gravity less than 1.00 would be lighter than water, and any compound with a specific gravity greater than 1.00 would be heavier than water. In general, petroleum-based compounds are lighter than water, and chlorinated compounds are heavier than water. Spills from gasoline stations and oil pipelines will usually be found on top of the aquifer. Of course, the new unleaded gasolines use several types of organic ring (benzene being the basic structure) compounds to enhance the octane rating.

**TABLE 3-2. Specific Gravity of Specific Organic Compounds.**

| COMPOUND | SPECIFIC GRAVITY |
|---|---|
| 1. Acetone | 0.79 |
| 2. Benzene | 0.88 |
| 3. Carbon tetrachloride | 1.59 |
| 4. Chloroform | 1.48 |
| 5. Methylene chloride | 1.33 |
| 6. Chloro benzene | 1.11 |
| 7. Ethyl benzene | 0.87 |
| 8. Hexachloro benzene | 1.60 |
| 9. Ethylene chloride | 1.24 |
| 10. 1, 1, 1-Trichloroethane | 1.34 |
| 11. 1, 1, 2-Trichloroethane | 1.44 |
| 12. Trichloroethylene | 1.46 |
| 13. Tetrachloroethylene | 1.62 |
| 14. Phenol | 1.07 |
| 15. 2-Chlorophenol | 1.26 |
| 16. Pentachlorophenol | 1.98 |
| 17. Toluene | 0.87 |
| 18. Methyl ethyl ketone | 0.81 |
| 19. Naphthalene | 1.03 |
| 20. Vinyl chloride | 0.91 |

[a]From Weast, Robert. "Handbook of Chemistry and Physics," 60th ed.: CRC Press, Inc., 1979, 1980.

These ring compounds are relatively soluble in water (Table 3–1). Although the main part of the gasoline will float, the soluble components will enter the water. This is also true for the alcohol part of the "gasohol" gasolines.

Spills of chlorinated hydrocarbons usually come from storage tanks where the compounds are used as solvents. Chlorinated organics are soluble in very low concentrations. The more chlorine substitutions, the less soluble the compound (Table 3–1). They will show up in the part-per-billion range in the aquifer. However, the threshold taste level for these compounds is also very low and people do notice their presence. More importantly, chlorinated hydrocarbons have been shown to cause cancer in laboratory animals. There has been a public outcry about any organics entering the potable water system. The main part of the compounds will not enter the aquifer, but instead will continue down through it until stopped by an impermeable soil layer. As with an impermeable layer in the unsaturated zone, the chlorinated compound will then spread like a pancake.

In both cases, the pure compound must be removed or it will be a continuous source of material to contaminate the aquifer. The designer should stress removal of these contaminants as pure compounds, which will be the fastest and least expensive method. By far the easiest compounds to recover are those that float. A floating material can be removed from an aquifer with relative ease because the water in the aquifer can be used to direct the movement of the floating compound. Figure 3-1 shows a floating material on the aquifer. If the water level is lowered in a particular place, the organic compound will continue its original downward path.

Figure 3-2 shows a well being placed in the middle of the spill area. Water is removed, and the water level in the immediate vicinity decreases. The nonsoluble compound will follow the lowering of the groundwater and concentrate at the lowest point—the well. The well can then also be used to remove the floating material. Water should be

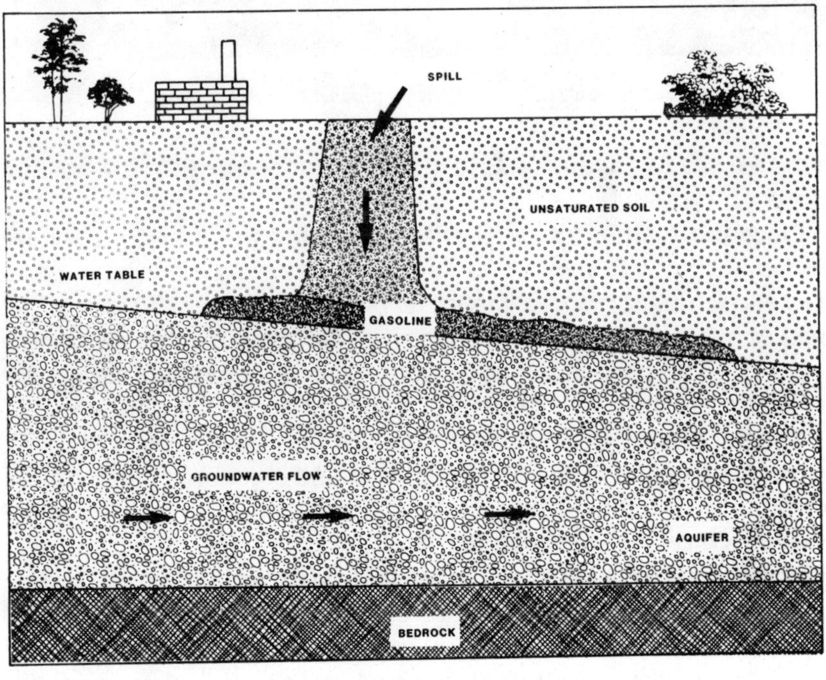


Figure 3-1. Petroleum product floating on top of an aquifer.

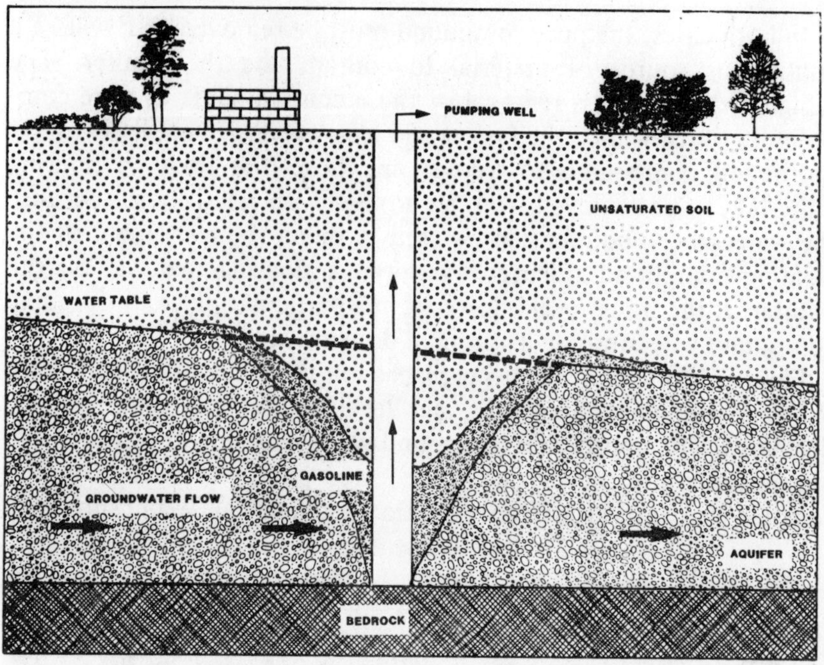


Figure 3–2. Collection of petroleum product by water drawdown.

removed at a fast enough rate to affect the entire spread of the floating material.

Once the floating material is accumulating in the well, there are several methods to remove it. The first is lowering a bucket into the well and letting the floating material pour over the top. The lip of the bucket is kept above the water level so that only the contaminant is removed. This process is time consuming and requires high manpower input. Several manufacturers have made this process more automatic.

The buckets have been weighted so that they float in water and sink in anything lighter than water. The same unit contains a pump for the water to maintain a depression and a timer to raise and lower the bucket. Once the well is drilled and the equipment is set up, only periodic visits to check the equipment are required.

Figures 3–3 and 3–4 show an even more advanced method of removal. This system, once again, uses a pump to maintain the water level. An oil/water separator is lowered into the well. A screen in the

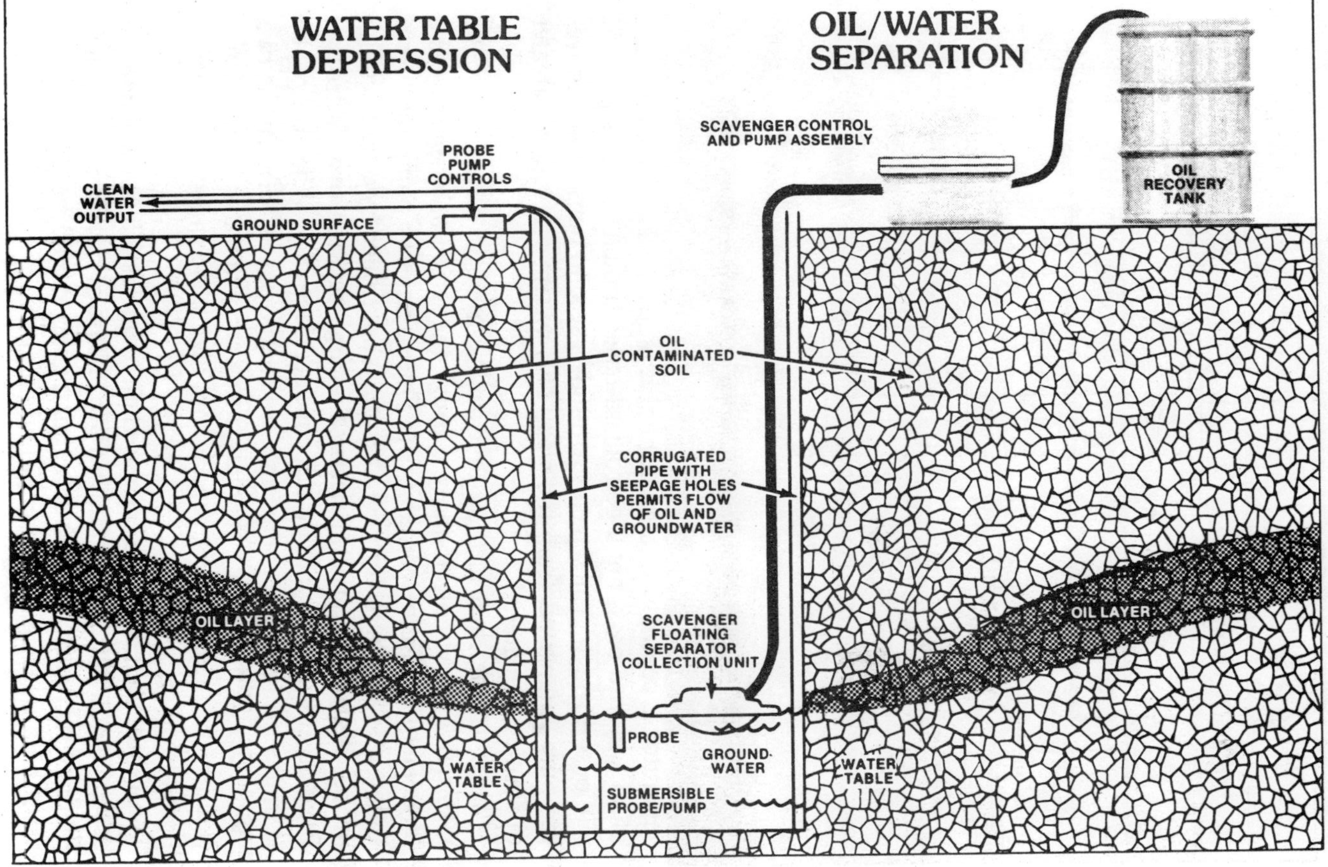


Figure 3-3. Oil/water separation in the well. *(Courtesy of Oil Recovery Systems, Inc.)*

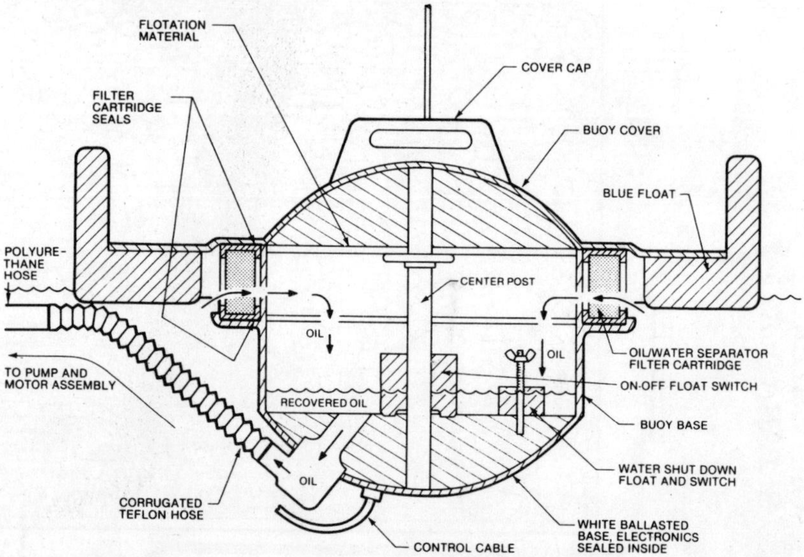


Figure 3-4. Oil/water separator. *(Courtesy of Oil Recovery Systems, Inc.)*

separator is coated with a hydrophobic material. The screen will not allow water to pass, but a petroleum-based product such as gasoline will pass through. The screen surrounds a second pump that removes the pure compound from the well. This system can remove greater quantities of material.

There is one final method that can be used, in a well, when the material is too viscous to move through the screen or when there is a large amount of pure contaminant. Once again two pumps are used. In this case, the pure-compound pump is controlled by a conductivity probe. Water can carry an electrical charge between two electrodes. Pure organic compounds do not conduct electricity. When the probe is immersed in a pure compound, no charge passes between the two electrodes of the probe. The unit then knows that it is in pure compounds and turns on the second pump. Figure 3-5 shows a conductivity removal method.

All three methods have been used on many installations. There are several manufacturers that have a broad base of experience with each of these methods. One of these companies should be contacted to assist on a floating, pure-compound recovery.

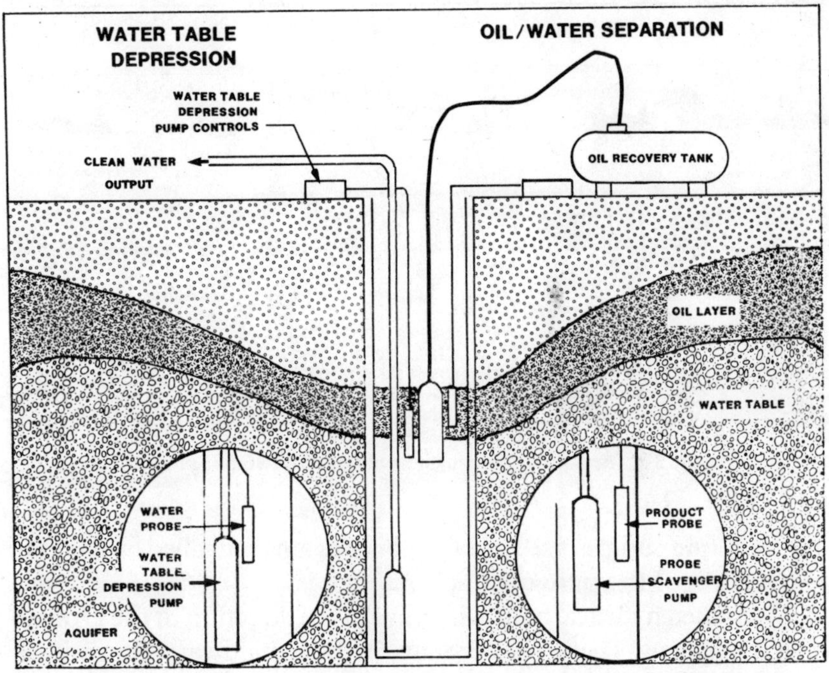


Figure 3-5. Oil removal with a product pump controlled by a product probe.

One other way to remove pure compounds that float is to dig a trench in the path of the groundwater movement, Figure 3-6. This technique works best for high water tables and other conditions where the spill stays near the surface (4). As the groundwater enters the trench, the floating organic remains on top of the water. However, once in the trench, the floating material can be removed by skimmers or other standard surface removal devices. The water level in the ditch can be lowered to increase the rate of movement of the floating material. Once again, the water in the aquifer is used to control the movement of the pure compound.

Pure compounds that sink cannot use the water in the aquifer as readily to control their movement. This makes their recovery very difficult. The first problem is to find the pool of pure material. The test well must drill directly into the material. Lowering the water level will not bring the material to the well. Once the material is found, the only way to control the removal is to use the conductivity method. In this

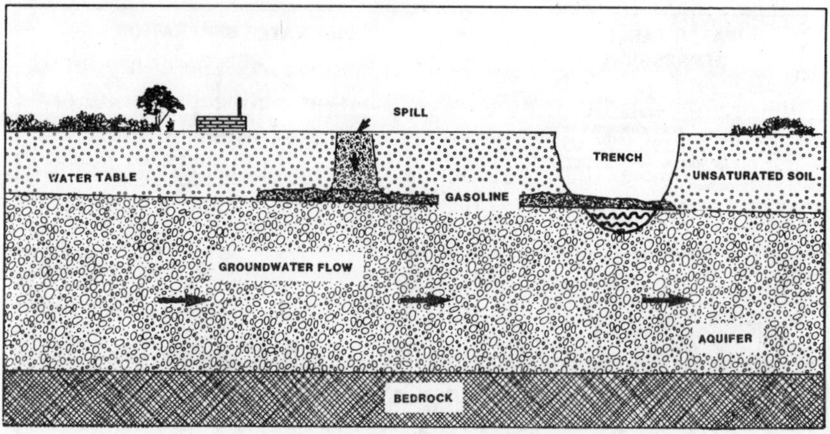


Figure 3-6. Removal of floating material by an intercepting trench.

case, only one pump is used, and it is once again controlled by the nonconductance of the pure organic compounds.

Chlorinated hydrocarbons will move in the direction of the groundwater movement. Therefore, a change in the groundwater flow will affect the movement of the chlorinated compounds. The effect will be limited. These compounds are found on the bottom of the aquifer. Gravity will not force these compounds to gather in the zone of depression. Wells can be used to force the chlorinated hydrocarbons in a particular direction, but the process will not be as efficient as with floating material.

In summary, when a pure compound is released to the ground, the design engineer must consider several important factors in order to clean up the release. The engineer must first determine if the material has reached the aquifer and must then discover if the material has stayed on top of the aquifer, dissolved into the water of the aquifer, or passed down through the aquifer. If the material has not dissolved into the water, then one of the pure-compound treatment methods should be used to recover as much of the material as possible. The rest of this chapter and Chapter 4 will cover treatment technology for the material that dissolves into the water of the aquifer and the contaminants that are left in the unsaturated zone.

## AIR STRIPPING

Air stripping is a process employed to remove volatile compounds (usually organics) from groundwater. The basic concept behind any air-stripping facility is to bring the contaminated water into intimate contact with air, so that the volatile compounds undergo a phase change (from liquid phase to vapor phase). The air will then carry off the contamination, leaving the water free from these compounds.

This aeration can be accomplished in several ways, depending upon the nature and duration of the cleanup project. These methods include

1. Aeration tanks
2. Cascade aerators
3. Spray basins
4. Packed towers

Of these four methods, the first two are generally not employed, due to economic considerations. Aeration tanks, in which compressed air is bubbled through a tank of water, generally cannot provide enough air to economically remove a high percentage of organics (5). Cascade aerators, on the other hand, have no operating cost associated with the air supply, since they are open to the atmosphere. They simply do not provide adequate air-water contact. Removal efficiencies obtained using cascade aerators are much less than those obtained with packed towers.

Spray basins have found use in several groundwater cleanup cases. In these systems, a piping grid is laid out over the area of a basin (either earthwork or concrete) and spray nozzles are placed evenly throughout the area to spray the contaminated water into the air in very fine droplets. The water is then either collected in the basin and pumped off or, in the case of an earthwork recharge basin, allowed to percolate back into the ground. This method is often used to flush out an area of contaminated soil. The advantage of this system is its extremely low capital cost, which makes it ideal for a cleanup of a temporary nature. The disadvantages are that large tracts of land are required and that neighboring properties may be affected by wind-driven mists or, in the winter, ice crystals. One installation in the northern United States solved this problem by erecting a 40-ft-high tarp along a property line, but this is a stopgap solution at best. Extra pumping costs may also be incurred in providing adequate pressure to the nozzles.

A packed tower is the method of air stripping that has found the most acceptance for both potable water purification and remedial work on groundwater contamination. In fact, the terms *packed tower* and *air stripper* are often used interchangeably.

Packed towers utilize a countercurrent flow scheme: Water enters at the top of the tower and flows downward through the packing, while the airstream flows upward, picking up the volatile compounds as it goes. The water is collected at the bottom and pumped to its final destination, while the air exits at the top of the tower and is dispersed, along with the volatiles, into the atmosphere (Figure 3–7).

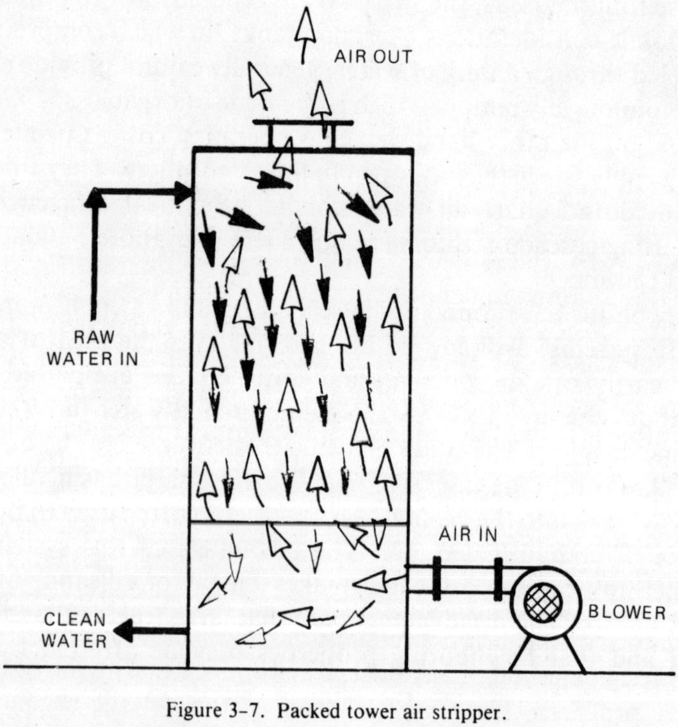


Figure 3–7. Packed tower air stripper.

## Design of Packed Towers

When sizing a packed column, the design engineer has three basic variables to define: (a) tower height, (b) tower diameter, and (c) air:water ratio. Although these variables are dependent upon each other (i.e., a change in air:water ratio may allow, or require, a change in packing height), the following basic relationships are helpful in preliminary sizing estimates:

*Tower diameter* is most strongly a function of flow rate; the cross-sectional area of a tower is proportional to the flow rate.

*Tower height* is most strongly a function of removal efficiency required. This relationship follows the law of diminishing returns; each incremental foot of packing added will remove a smaller amount of contaminant than the previous foot.

*Air:water ratio* is a function of the contaminant being removed. The more volatile a substance is, the less the amount of air that is required to strip it.

A fourth variable that controls the stripping process, temperature, may also be changed through the use of preheaters of the water stream or injection of steam directly into the tower. Stripping will occur at a much higher rate at elevated temperatures, and some compounds that are barely volatile at ambient temperatures can be totally removed by raising the column temperature into the 140–180°F range (6). The use of high-temperature strippers or steam stripping is generally limited to hazardous waste site cleanups of short duration, due to the high operating costs associated with heating the process streams. Ambient-temperature strippers do not experience any change in operating performance between winter and summer operation, although air temperatures may fluctuate over a 100°F range. This is because the groundwater temperature stays at a near constant 50–55°F year-round, and the thermal mass of the water is much greater than the thermal mass of the air in a stripper. Thus, the actual operating temperature of the stripper remains fairly constant.

The design method of any packed column starts with the basic mass transfer process. The rate of transfer of the volatile organic compound will be a function of the driving force (the concentration gradient between water and air) and the air-water interface area. Different compounds will be transferred at different rates, depending on the Henry's law constant of the particular compound. Henry's law constant is the ratio of the partial pressure of a compound in air to the mole fraction of the compound in water at equilibrium. Compounds with a high Henry's law constant have a greater concentration in air when an air-water system is in equilibrium. These compounds undergo a phase change from liquid to vapor quite easily and hence are easily stripped. Compounds with low Henry's constants, on the other hand, are more hydrophilic and are more difficult to strip. Table 3–3 provides the Henry's law constants for 20 organic compounds.

The equation used to determine the operating parameters of a tower is

$$Z_t = \left(\frac{L}{K_L a} \bullet \frac{R}{R-1}\right) \ln \left[\frac{(C_{in}/C_{eff})(R-1)+1}{R}\right] \tag{3-2}$$

where:

$R = H \bullet G/L \bullet P_t$ (stripping factor)

and:

$H$ = Henry's law constant (atm m³ $H_2O$ m⁻³ air)
$G$ = gas loading rate ($m^3\ m^{-2}\ sec^{-1}$)
$L$ = liquid loading rate ($m^3\ m^{-2}\ sec^{-1}$)
$P_t$ = operating pressure (= 1 atm)
$Z_t$ = packing height (m)
$C_{in}$ = influent concentration (consistent units)
$C_{eff}$ = effluent concentration (consistent units)
$K_L a$ = mass transfer coefficient ($sec^{-1}$)

The key variables to define in Equation 3–2 are Henry's law constant ($H$) and the mass transfer coefficient ($K_L a$). A rough estimation of Henry's law constant is usually sufficient for design work (Table 3–3). This is fortunate, since published values of Henry's constants can vary

**TABLE 3-3. Henry's Law Constants for Specific Organic Compounds.**

| COMPOUND | HENRY'S LAW CONSTANT[a] (ATM •$M^3$ WATER/$M^3$ AIR) |
|---|---|
| 1. Acetone | 1,300 |
| 2. Benzene | 240 |
| 3. Carbon tetrachloride | 1,300 |
| 4. Chloroform | 200 |
| 5. Methylene chloride | 140 |
| 6. Chloro benzene | 230 |
| 7. Ethyl benzene | 350 |
| 8. Hexachloro benzene | 33 |
| 9. Ethylene chloride | 60 |
| 10. 1, 1, 1-Trichloroethane | 220 |
| 11. 1, 1, 2-Trichloroethane | 48 |
| 12. Trichloroethylene | 450 |
| 13. Tetrachloroethylene | 1,100 |
| 14. Phenol | 0.016 |
| 15. 2-Chlorophenol | 1.74 |
| 16. Pentachlorophenol | 0.13 |
| 17. Toluene | 330 |
| 18. Methyl ethyl ketone | 0.99 |
| 19. Naphthalene | 22 |
| 20. Vinyl chloride | 390,000 |

[a]From Sullivan, Kevin. Summary of Henry's Law Constants. Used by Hydro Group. August 1984.

by more than an order of magnitude. Obviously, more theoretical research needs to be done in this area.

The mass transfer coefficient is a function of tower design and type of packing, and it is a good expression of the overall efficiency of the tower. Accurate quantification of this coefficient is very important, since, as can be seen from Equation 3–2, there is an inverse relationship between tower height and $K_La$. Thus, a 25% error in a $K_La$ value might add 25% to the height of the tower, resulting in increased costs due to additional shell material and packing required.

Because of this relationship, it is good engineering practice to select a mass transfer coefficient based on some type of field data, such as results of a pilot test or operating data from a similar installation on a similar water supply. A pilot study on the actual water to be treated is best, since chemical characteristics (such as the type and amount of pollution) will vary from source to source and may affect the stripping process. Pilot testing can be simple and inexpensive; usually, enough data for design purposes can be obtained in a single day of testing. The testing should cover a range of possible liquid loading rates and air:water ratios, as the $K_L a$ will increase with both of these factors. The engineer selects a Henry's law constant value for the compound to be tested, then varies the gas and liquid loading rate and measures the resulting effluent concentration for a constant influent concentration. Equation 3–2 can then be solved for the mass transfer coefficient.

When Henry's law constant is known and a $K_L a$ determined for the range of water loading rates and air:water ratios of interest, then Equation 3–2 may be solved for different combinations of packing height and liquid loading rates. The design engineer can use these data to generate various tower configurations that will provide the required removal efficiencies. Some of these options can be eliminated due to site-specific restraints, such as a maximum allowable height. After these are removed, an estimate of capital costs and operating costs should be made for each tower, and a final tower design should be selected.

Cost of treatment can vary widely for packed towers, depending on the removal efficiency required and the compounds involved. The costs range from approximately \$0.04/1000 gal for a large unit treating a fairly volatile organic to \$0.17/1000 gal for a stripper that requires a very high air:water ratio and hence has high operating costs due to the energy requirements of the blowers.

## Column Components

The major components of a stripping tower are the tower shell, tower internals, packing, and air delivery systems (Figure 3–8). The tower shell is usually cylindrical, for strength, for ease of fabrication, and to avoid any corners that might induce channeling of the air or water. The tower must be built to withstand all applicable wind, snow, and earthquake loads for the area in which it is being installed, and additionally must be able to support the combined weight of the tower internals, the packing, and the water held up in the tower. However, the tower does not need to be designed to support a full column of water, since it will never be entirely filled with water during operation. Air openings at the bottom of the tower will allow water to drain out in

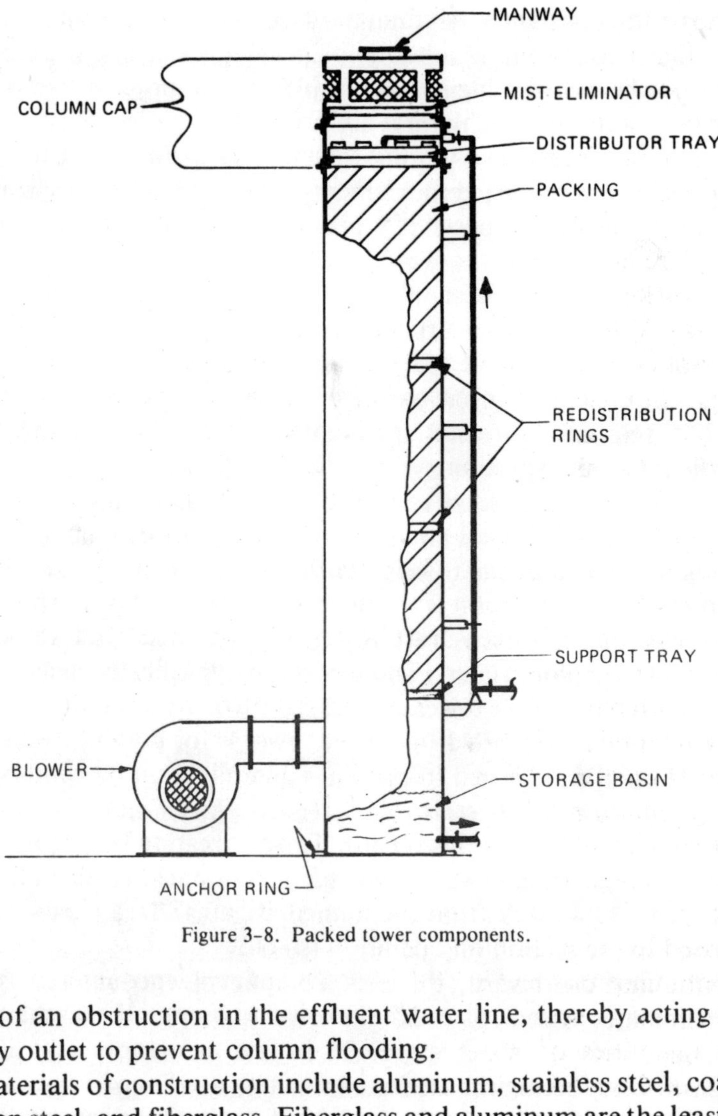


Figure 3-8. Packed tower components.

case of an obstruction in the effluent water line, thereby acting as a safety outlet to prevent column flooding.

Materials of construction include aluminum, stainless steel, coated carbon steel, and fiberglass. Fiberglass and aluminum are the least expensive, with stainless steel and coated carbon steel being slightly higher in cost for most cases. However, it is important to remember that metal prices can fluctuate rapidly, so all applicable materials should be considered at the time of tower design. Costs aside, the relative advantages of thc various materials hinge on their strength and corrosion resistance. Aluminum has excellent structural properties, is lightweight, and is suitable for potable water applications. It should not be used on highly acidic (pH $< 4.5$) process streams or where large amounts of chlorides are present. Fiberglass towers offer good corrosion resistance in most chemical environments (the hand lay-up method is preferred to filament-wound columns for chemical resistance) but are comparatively brittle, making it difficult to construct tall towers subject to wind loads.

The various stainless steel alloys offer a wide range of corrosion resistance as well as good structural properties, but many alloys are often not readily available at less than premium prices. Carbon steel with an epoxy coating offers corrosion resistance and strength at low cost but requires increased maintenance costs for painting and periodic internal inspection.

The tower internals serve to insure that the mass transfer process takes place under optimal conditions, at the most economical cost. Starting at the top of the tower, the first component that requires a design engineer's attention is the air exhaust ports. (For purposes of this discussion, a forced-draft tower will be used. Induced-draft towers will be explored later.) These ports are typically located around the circumference of the tower and are sized to permit the air to escape with a minimum pressure drop. If the tower is for potable water, the outlets should be screened to prevent contamination by wind-borne material entering the tower; towers screened with 24-mesh screen have reported no problems in this regard. Towers treating large quantities of heavily contaminated water may require tall stacks to direct the exhaust air up and away from the immediate area. This is more fully discussed in the section on operation (p. 60).

Continuing downward, the next component encountered is the mist-eliminator system, placed in the tower to prevent the discharge of large quantities of water entrained in the airstream. This is accomplished by forcing the airstream through a series of bends to impinge the water droplets on the surface of the mist eliminator. When the droplets grow to a large enough size, they fall back into the tower. Mist eliminators come in two broad categories: cheveron-type and pads. The cheveron-type eliminators are made up of bent metal plates placed next to each other such that the air is forced to zigzag through, impinging the water on the metal plates. The pad types are composed of filaments loosely bundled or woven into pads ranging from 4 in. to 12 in. thick. As the air flows through the pad, the water droplets are deposited on the filaments, which are usually made of polypropylene or stainless steel. The pad-type mist eliminators can provide greater mist removal, especially at higher airflow rates. They do, however, have a higher pressure drop.

Water is introduced into the tower by means of a distributor, which insures that the water is evenly distributed across the surface of the packing, while allowing for smooth, unimpeded airflow upward to the top of the tower. The distributors fall into four general categories: distributor trays, trough-and-weir arrangements, header-lateral piping, and spray nozzles. The header-lateral and trough-and-weir systems rely on the same basic concept—dividing the flow into successively smaller streams. The major drawback of these systems is the difficulty in assuring even water distribution, a factor that is critical to efficient tower operation. Weir systems have certain "blind" spots under the troughs where water does not fall, and header-lateral systems are notorious for unequal flow in different laterals, depending on their location. However, these systems do find use where large airflows are required, since they provide a large open area through which the air can pass.

Orifice-type distributor trays avoid these problems of unequal distribution. The trays are designed to keep a standing head of water in them, thereby assuring that an equal pressure, and hence an equal flow, will be maintained at each orifice. Air stacks are provided to allow gas flow through the tray. These trays do an effective job and are generally less expensive to fabricate than other distributor systems.

Spray nozzles have found use in a number of applications. Their major advantage is that they immediately break up the water flow into droplets, thereby enhancing the mass transfer process. The major problems are the increased water pressure required to operate them (which increases pumping costs), the extra tower space required to allow for their use, and clogging of the nozzle, especially in turbid waters.

Below the distributor lies the packing, which is held up by a packing support plate. This plate must be structurally capable of supporting not only the weight of the packing but also the weight of any water present in the packed bed. At the same time, it must have enough open area to avoid flooding, a condition that results when the water flow downward through the tower is significantly impeded by the gas flow. When the water flow is restricted, a head of water builds up until the water weighs enough to force its way through the plate. This leads to an unstable "burping" action, where first water and then air are alternately passed through the plate, resulting in decreased removal efficiencies in the tower.

For most water treatment applications, a fiberglass reinforced plastic (FRP) grating will provide adequate open area to prevent flooding. In designs with very high liquid and gas loading rates, a gas injection plate is sometimes employed. These plates have a wavy appearance, which provides more open area than is possible with a flat plate. Air is "injected" through the top peaks of the plate. These plates are usually fabricated out of stainless steel, and they are more expensive than FRP grating.

The design of the base of the tower will vary with system configuration; an integral clearwell may be supplied as part of the tower, or the water may flow by gravity to discharge in a stream or sewer. Whatever the configuration, it is imperative that a water seal be provided in the discharge line, to prevent short-circuiting of the tower by the air blowing out the discharge line. A hydraulic analysis of the discharge should also be performed to insure that the water will not back up in the tower, possibly flooding the air blowers.

The single most important component selection for the design engineer is the tower packing (Figure 3–9). The ideal tower packing will provide a large surface area for the air and water to interact, and it will also create turbulence in the water stream to constantly expose fresh water surfaces to the air. The packing should have a large void area to minimize the pressure drop through the tower. Secondary considerations for a packing include weight, corrosion resistance, ability to maintain a uniform liquid flow, and, of course, price.

Unfortunately, there is no single measure to determine the best tower packing. Measurements such as surface area/unit volume can be misleading because the surface area of the packing is not the same as the air-water interface area. Many of the packings with wide, sweeping surfaces may have only one side of their area wetted, essentially wasting the other half by leaving it dry. Other packings, because

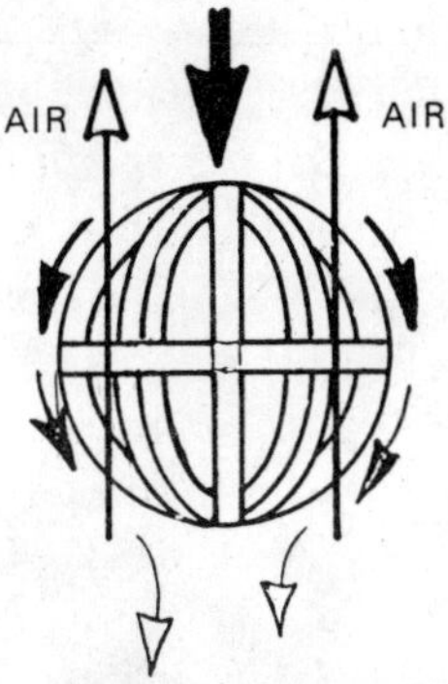


Figure 3-9. Tower packing.

of their configuration, may cause channeling of the water, reducing the air-water contact area. For this reason, comparison pilot-scale tests using various packings are the only valid method of evaluation.

One of the great improvements in mass transfer technology has been the introduction of inexpensive plastic packings in recent years. The use of injection molding has allowed the creation of packings much more suited to the dual goals of maximum mass transfer surface and minimum pressure drop. Early packings, such as saddles and rashing rings, were limited to fairly simple shapes by the nature of their production process, usually in metals or ceramics. The packings shown in Figure 3-10 are all made of polypropylene, and their complex shapes assure a large void area to minimize pressure drop. Pilot testings of these packings have shown much better results than those obtainable with the older packings.

Polypropylene packing has several other benefits. It is chemically inert and will not degrade when exposed to most chemicals encountered in groundwater contamination. It is very inexpensive; an equal volume of ceramic packing may cost 8-10 times as much. Finally, it is lightweight and strong, allowing greater packed bed depths without crushing the packing at the base of the tower. This strength also allows the packing to be dumped into a tower without damage; ceramic packing must be loaded into a tower filled with water to

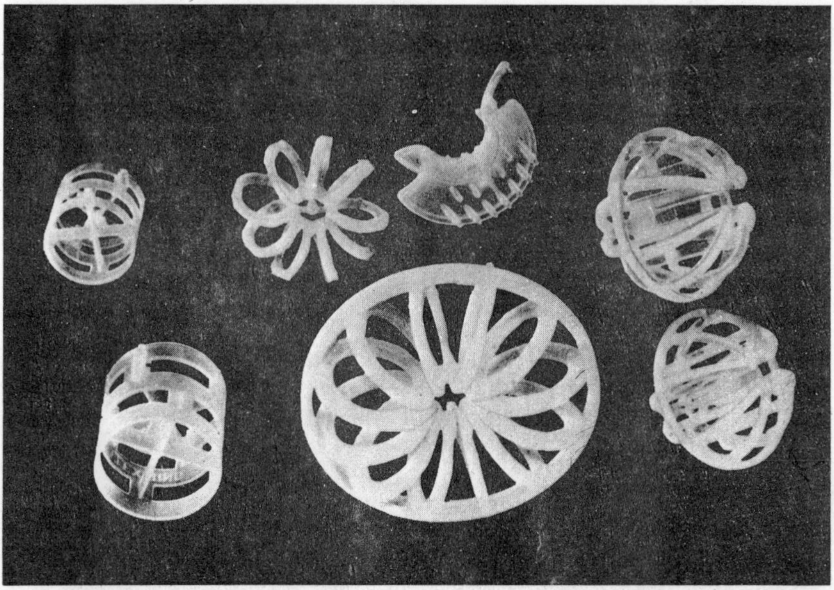

Figure 3-10. Examples of plastic tower packing. *(Courtesy of Hydro Group.)*

cushion its fall. Designing a tower to withstand such hydraulic loading adds to the cost of installation.

The packings shown in Figure 3-10 all fall into one category: the random, dumped packings. These packings are simply dumped into the tower and allowed to rest in whatever configuration they land. The other broad category of tower packing is the stacked packings. These packings are physically stacked in the tower and configured in such a way to assure even water redistribution. These packings are usually not used in water treatment, due to their high capital cost and the additional labor cost of their installation. Their single advantage is that somewhat shorter packed depths are theoretically required. Test results with these packings have yet to clearly demonstrate this fact.

The final component of an air-stripping system is the delivery system. Usually a forced-draft blower is provided at the base of the tower or housed nearby in a building if sound levels are a concern. However, with proper tower design and selection of a minimum air:water ratio, the size (and hence the noise) of a blower can be kept to a minimum. Sound mufflers are available for insertion over the air inlet if desired, but these result in an increased pressure drop. A complete packed tower is shown in Figure 3-11.

The alternative to a forced-draft tower is an induced-draft system, where the air is drawn through the tower by a blower. These systems are limited to somewhat lower pressure drops, but often find application where the gases being discharged undergo further treatment before their release.

Figure 3-11. Example of a packed tower air stripper. *(Courtesy of Hydro Group.)*

## Operation

Once an air stripper is installed, its operation is a very simple matter. The air delivery system is controlled by the water supply pumps, so that whenever water is introduced to the tower, the blowers are turned on. Maintenance is minimal, and periodic inspection of the bed is all that is required internally.

Several concerns about the operation of a stripping system are frequently raised. The first and largest concern is the air pollution that is created by discharging the volatile organics to the atmosphere. The contamination is not destroyed in a mass transfer process; it is merely changed in form. However, two factors mitigate the effects of these atmospheric discharges. The first is the dilution that takes place in the tower before the vapors are emitted. Air:water ratios commonly employed range from 25:1 to 250:1. Thus, the pollutant is diluted by a similar factor when it is transferred into the air. On top of this dilution, there is the natural dilution that occurs as soon as the airstream is dispersed into the atmosphere.

The second factor is that many compounds, such as TCE and PCE, will break down in the atmosphere under the effects of the sun's natural radiation. TCE, for example, has a half-life of approximately a day and a half in the atmosphere. The history of environmental treatment in the United States, however, has been toward destruction or final disposal, not switching the pollutant from one medium to another. Already several states require that all air discharges from stripping towers be treated before being released to the atmosphere.

In these states and in circumstances where the total discharge to the atmosphere is too high, the exhaust gases are usually treated by passing them through an activated-carbon bed. At first glance, this configuration appears superfluous, since liquid-phase carbon could treat the water directly. However, this system may save on carbon costs, because vapor-phase carbon can often hold more contaminants before it becomes saturated. There will also be fewer chemicals in the vapor stream competing for the available pore space, since many harmless compounds will remain in the liquid phase. Finally, a stripper allows removal of any desired amount of contaminant, from less than 50% up to 99.99%. Carbon removes close to 100% of the compound until breakthrough occurs. This "all-or-nothing" use of carbon can become very expensive.

Other air treatment techniques can be used. When the contaminant has some value, and sufficient quantity is present, recovery techniques may be employed. Another method of treatment that has been reported is the incineration of the gas stream (7). However, the capital and operating costs will normally prohibit this technique. Some work has been performed on catalytic oxidation of chlorinated hydrocarbons in the gas stream, but no detailed data have been reported. Air stripping is a relatively new treatment technique for organic compounds. More treatment methods for the gas stream can be expected in the future.

Other common concerns about stripping tower operation include freezing, bacterial growth, and plugging of the packing. Freezing can be avoided by designing the tower and internals to be entirely self-draining, avoiding any interior ledges or pockets that could develop more than a thin coating of ice. Bacterial growth has not been a problem on columns in potable water plants that have been on line for over five years. Plugging may occur in some waters with very high iron content (usually greater than 3–5 ppm); the iron is oxidized in the tower and will precipitate out onto the packing. Research is currently under way to assess both the extent of the problem and the possible methods of dealing with it.

In all, air stripping is probably the fastest-growing method for treatment of groundwater. The combination of low cost, easy operation, and the wide variety of compounds that can be removed from groundwater make stripping the first choice for low-concentration streams. Even in cases when the air discharge must be treated, air stripping has still been found to be the least costly method.

## CARBON ADSORPTION

The use of carbon for its adsorptive qualities was known as early as 1550 B.C., when charcoal was utilized in the purification of medicines. In the field of water treatment, both ancient Egyptians and eighteenth-century sailors utilized charcoal-lined vessels to provide for clean drinking water.

The use of carbon as a water treatment process, though, traces its roots to London in the 1860s, when some of the residents had their drinking water filtered through animal charcoal to remove tastes and odor. Granular-carbon filters were introduced in the 1930s for the ultrapure water in the food and beverage industry.

Following World War II, coal was utilized to produce high-activity, hard granular carbons on a commercial scale, leading to the widespread use that granular activated carbon has today.

The use of activated carbon to remove taste and odors from drinking water supplies is now an established technology. Since the introduction of activated carbon on a commercial scale, industry has also taken advantage of its unique ability to adsorb a variety of organic compounds by utilizing it in product purification as well as in water and wastewater treatment.

Based on its history and unique properties, activated carbon has now become a proven technology for removal of synthetic organic contaminants from groundwater, even though these contaminants do not exhibit traditional taste and odor characteristics and may be present in trace-level concentrations rather than the high levels found in wastewaters.

## Concepts of Adsorption

Adsorption is a natural process in which molecules of a liquid or gas are attracted to and then held at the surface of a solid. Physical adsorption is the attraction caused by the surface tension of a solid that causes molecules to be held at the surface of the solid. Chemical adsorption involves actual chemical bonding at the surface solid. Physical adsorption is reversible if sufficient energy is added to overcome the attractive forces, whereas chemical adsorption is not a reversible reaction.

Adsorption on activated carbon is of a physical nature. What makes this material such an excellent adsorbent is the large degree of surface area contained within the carbon particle that is accessible for the adsorption process. Surface areas of granular carbons range up to 1400 $m^2$ per gram of material. As the surface area of the activated carbon is internal to the material, crushing the granular material will not increase its surface area. Even in its crushed or powdered state, activated carbon still retains internal surface area, making it an effective adsorbent.

Granular activated carbon has an internal structure that enables it to more fully utilize its surface area (Figure 3–12). Macropores, or larger openings with diameters greater than 1000 Å, provide an entrance into the interior structure of the carbon particle. Although some adsorption may take place along these pores, they serve to conduct the molecules to the micropores, pores with diameters less than 1000 Å, where adsorption takes place. Many pores are small enough only to contain small molecules, so the effective surface area for adsorption for a particular species depends upon its size and the available surface area of the pores it can enter. Capacity of a particular grade of carbon, therefore, may vary for different species, so standard tests have been developed to identify capacities. These tests may utilize iodine molecules to identify small pores, and molasses to identify the macropore structure, for example. A wide variety of activated carbons are available, and properties such as surface area and pore size distribution will determine their applicability to any given situation.

The adsorption mechanism consists of three steps: (a) diffusion of the molecules through the liquid phase to the carbon particle, (b) diffusion of the molecules through the macropores to the adsorption site, and (c) adsorption of the molecule to the surface. The characteristics of the molecule will determine the rate of each step and, finally, the amount of time required for the entire adsorption process. Less soluble organics, for example, will diffuse rapidly to the granule, and large molecules will move slowly through the macropore structure. Generally, the chlorinated solvents found to be contaminating groundwater are amenable to activated-carbon adsorption, due to their low solubility and small molecular size, enabling effective use of the micropore adsorption area. It is important, however, for each contamination problem to be properly evaluated.

### Evaluation Procedures—Adsorption Isotherms

The first step in evaluation of activated-carbon adsorption for a specific evaluation is to assess its feasibility utilizing a liquid-phase adsorption isotherm test.

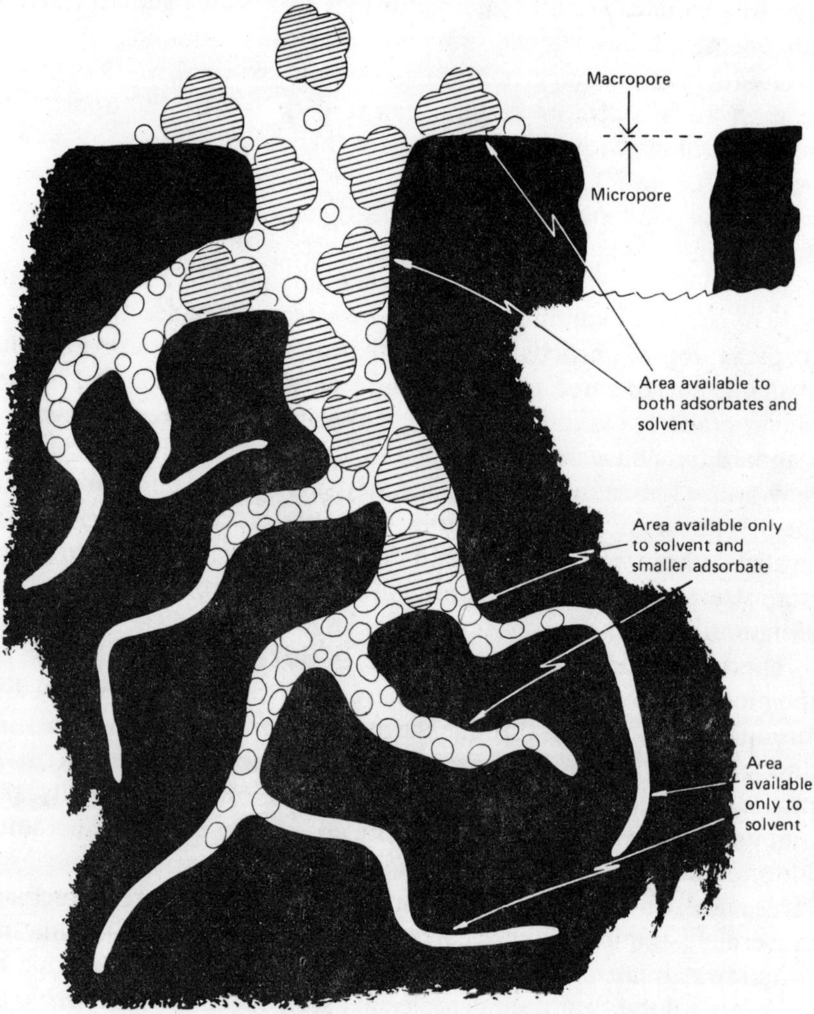


Figure 3-12. Internal structure of activated carbon. *(Courtesy of Calgon Carbon Corporation.)*

An adsorption isotherm is a batch test designed to demonstrate the degree to which a particular dissolved organic compound (adsorbate) is adsorbed on activated carbon (adsorbent). The data generated show the distribution of adsorbate between the adsorbent and solution phases at various adsorbate concentrations. From the data, a plot of the amount of impurity remaining in solution at constant temperature can be generated. For a single adsorbate, a straight-line plot (on log-log paper) can be obtained using the empirical Freundlich equation:

$$\frac{x}{m} = kc^{1/n} \quad \text{or} \quad \log \frac{x}{m} = \log k + \frac{1}{n} \log c \qquad (3\text{–}3)$$

where:

$x$ = amount of contaminant adsorbed
$m$ = weight of carbon
$c$ = equilibrium concentration in solution after desorption
$k$ and $n$ are constants

For mixtures of adsorbates, a series of straight lines can be obtained. The presence of a nonadsorbable component will result in a curvature of the line when in combination with an adsorbable component, and in a vertical line when alone.

Data for generating this type of isotherm are obtained by treating fixed volumes of the water sample with a series of known weights of carbon. The carbon-liquid mixture is agitated for a fixed time at a constant temperature. After the carbon has been removed by filtration, the residual adsorbate concentration is then determined. The amount of organic adsorbed by the carbon ($x$) is divided by the weight of carbon in the sample ($m$) to give one value of $x/m$ for the isotherm. For contaminants that are volatile at ambient temperatures, the isotherm can be conducted in low-temperature baths to maintain the contaminant in the aqueous solution.

To estimate the capacity of carbon for the contaminant, then, one uses the $x/m$ value that corresponds to its influent concentration, $C_o$. This value, $(x/m)C_o$, represents the maximum amount of contaminant adsorbed per unit weight of carbon when the carbon is in equilibrium with the untreated contaminant concentration.

Table 3–4 presents the isotherm data from a trichloroethylene contaminant at 1600 ppb. The data are summarized in Figure 3–13.

As an example, assume that trichloroethylene was present in groundwater at 0.2 mg/liter. According to the isotherm, the equilibrium capacity is 19.5 mg trichloroethylene adsorbed per gram of carbon, or a capacity of about 2%. Therefore, the amount of carbon required would be (0.2 mg/liter)/(19.5 mg/g) = 0.01 g carbon per liter of water, or approximately 0.1 lb per 1000 gal treated. This capacity is based on allowing the activated carbon to reach equilibrium with trichloroethylene, an ideal condition not usually ob-

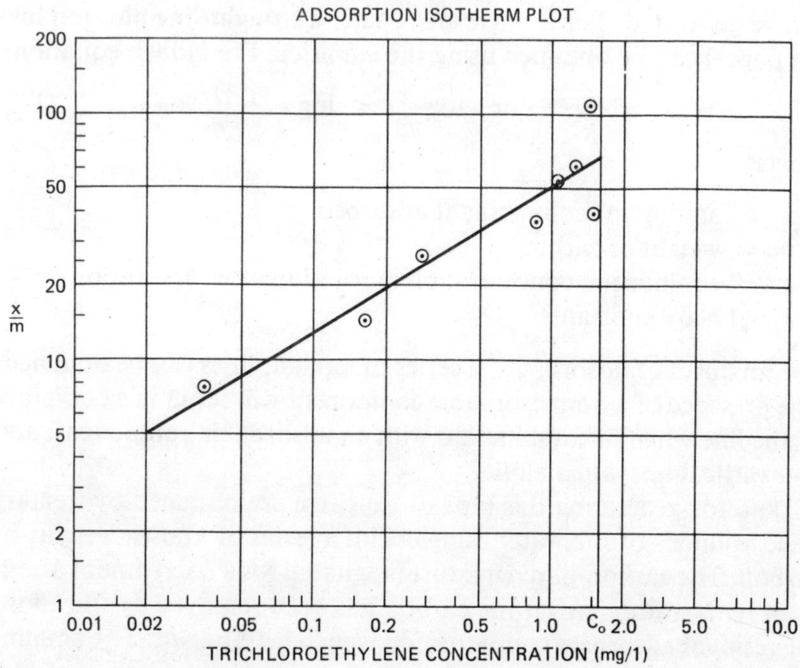


$$\left(\frac{x}{m}\right)_{C_o} = 66 \frac{mg}{g} \qquad \text{CORRELATION COEFFICIENT} = 0.928$$

Figure 3–13. Trichloroethylene adsorption isotherm plot.

tainable in practice. However, the isotherm evaluation does prove that carbon adsorption is feasible and should be evaluated further.

Table 3–5 summarizes the equilibrium adsorption capacities of 20 specific organic compounds at 500 ppb concentration of each compound.

### Evaluation Procedures—Dynamic Column Study

When an operating activated carbon adsorption system is designed, additional information that is not available from the adsorption isotherm must be obtained. The optimum operating capacity and contact time need to be determined to fix the adsorber size and optimum system configuration. The optimum contact time and mass transfer zone depend upon the rate at which the contaminant is adsorbed by the carbon, and they can only be determined by dynamic testing.

## TABLE 3-4. Carbon Adsorption Isotherm for Trichloroethylene.

| m CARBON (g) | c, TCE REMAINING (ppm) | (mg) | x TCE ADSORBED (mg) | x/m |
|---|---|---|---|---|
| Control | 1.600 | 0.800 | — | — |
| 0.0005 | 1.490 | 0.745 | 0.055 | 110.0 |
| 0.0010 | 1.520 | 0.760 | 0.040 | 40.0 |
| 0.0025 | 1.290 | 0.645 | 0.155 | 62.0 |
| 0.0050 | 1.060 | 0.530 | 0.270 | 54.0 |
| 0.010 | 0.860 | 0.430 | 0.370 | 37.0 |
| 0.025 | 0.285 | 0.143 | 0.657 | 26.3 |
| 0.050 | 0.165 | 0.083 | 0.717 | 14.3 |
| 0.100 | 0.035 | 0.018 | 0.782 | 7.8 |
| 0.250 | < 0.010 | — | 0.800 | — |
| 0.500 | < 0.010 | — | 0.800 | — |

Conditions:
- Type of carbon — Filtrasorb 300
- Temperature — ambient
- Sample volume — 500 ml
- Agitation time — 4.0 hr

## TABLE 3-5. Adsorption Capacity of Activated Carbon for Specific Organic Compounds.

| COMPOUND | MILLIGRAMS COMPOUND/GRAM CARBON (AT 500 PPB) | REF. |
|---|---|---|
| 1. Acetone | 43 | (3) |
| 2. Benzene | 80 | (3) |
| 3. Carbon tetrachloride | 6.2 | (5) |
| 4. Chloroform | 1.6 | (3) |
| 5. Methylene chloride | 0.8 | (8) |
| 6. Chloro benzene | 45 | (8) |
| 7. Ethyl benzene | 18 | (3) |
| 8. Hexachloro benzene | 42 | (8) |
| 9. Ethylene chloride | 2 | (5) |
| 10. 1, 1, 1-Trichloroethane | 2 | (5) |
| 11. 1, 1, 2-Trichloroethane | | |
| 12. Trichloroethylene | 18.2 | (5) |
| 13. Tetrachloroethylene | 34.5 | (5) |
| 14. Phenol | 161 | (3) |
| 15. 2-Chlorophenol | 38 | (8) |
| 16. Pentachlorophenol | 100 | (8) |
| 17. Toluene | 50 | (3) |
| 18. Methyl ethyl ketone | 94 | (3) |
| 19. Naphthalene | 5.6 | (8) |
| 20. Vinyl chloride | Trace | (8) |

The column test is conducted with a set of columns connected in series, as shown in Figure 3–14. Each column is filled with an amount of carbon calculated to provide superficial contact times of 15–60 min/column. The liquid rate to the column is usually in the range of 2 gpm/ft$^2$, although it may vary according to the test, as at this point the contact time is of more importance. The surface loading may be more important if there are suspended solids present and the activated carbon bed is to act as a filtering medium as well as an adsorption process.

Water is pumped through the column system and effluent samples are collected from each of the columns. The adsorption isotherm test should provide an estimate of how often testing should be done. The amount of the contaminant in the column effluent is plotted against the volume throughput of each column. The result is a series of curves, each curve representing a column. The successive curves also represent increasing contact times in a single bed. Figure 3–15 shows an example of a column study where each column represents 15 min of contact time. The curves obtained are termed breakthrough curves, as they represent the concentration or amount of contaminants present in the effluent (which have passed through the column unadsorbed).

The results of a dynamic column study are utilized to establish the design of an operating carbon-adsorption system. The first step is to establish the contact time required in the operating system. For each of the breakthrough curves established in the column study, a carbon usage rate can be calculated. This usage rate is the pounds of carbon required for a given volume of liquid to maintain the contaminant at a desired level in the effluent. The usage rate is calculated by dividing the amount of carbon on line by the volume treated up to when the desired effluent concentration is exceeded, or the break point of the breakthrough curve. The carbon usage rates can then be plotted for each contact time (column), as shown in Figure 3–16, and the optimum contact time determined as the point where increasing contact time obtains little improvement in carbon usage. The amount of carbon on line is then established by multiplying the contact time by the flow rate to obtain the volume of the carbon bed.

The next step is to consider whether only a single carbon adsorber is required or if a second unit in series would yield substantial benefits. Figure 3–17 shows the configuration of two breakthrough curves. The

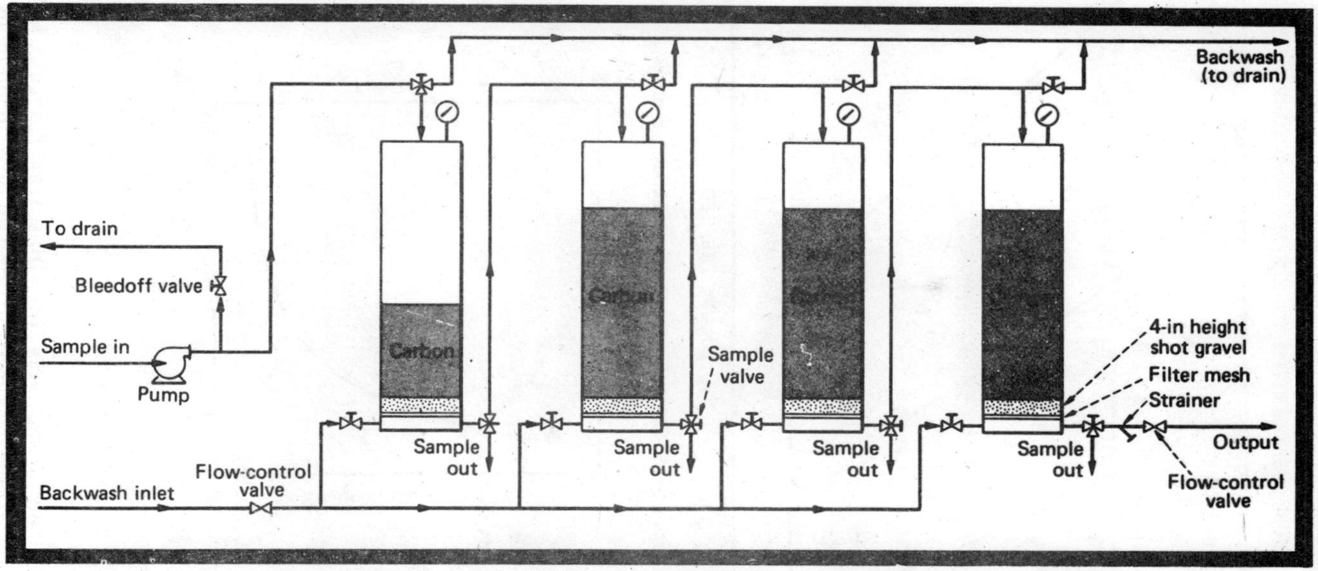


Figure 3–14. Laboratory series column adsorption test.

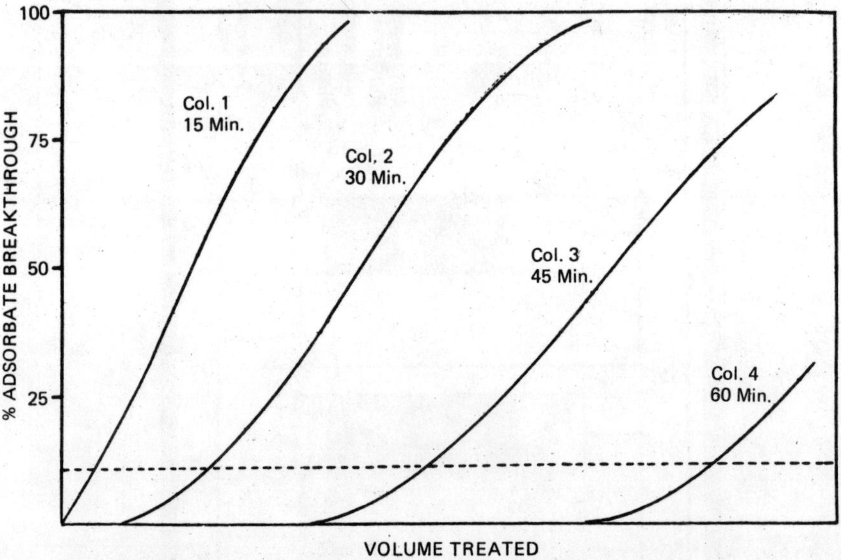


Figure 3-15. Column study results: breakthrough curves.

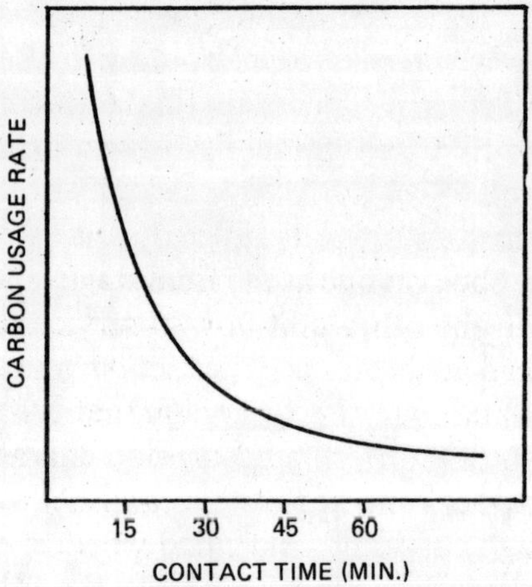


Figure 3-16. Optimum carbon contact time.

steep curve indicates a relatively short mass transfer zone. In this case, good utilization of the activated carbon can be realized in a single bed where the carbon is exchanged when the effluent concentration exceeds the desired level.

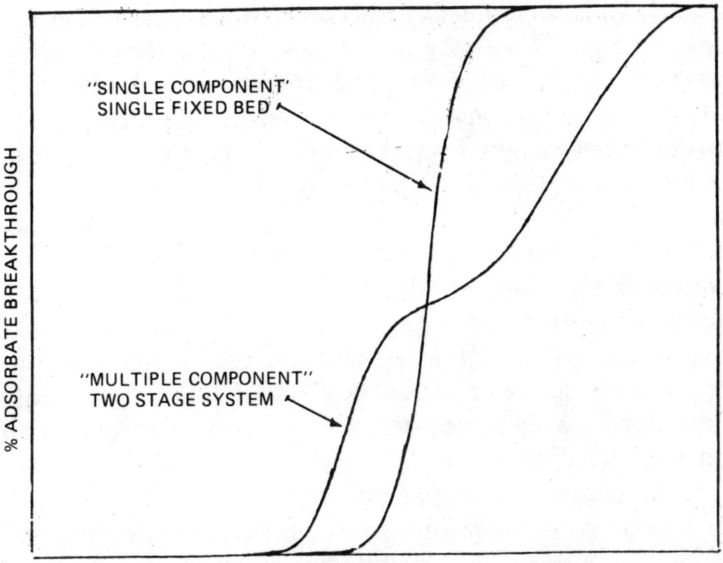


Figure 3–17. Mass transfer zones for two typical breakthrough curves.

The gradually sloping curve indicates a long mass transfer zone. For these instances, a staged system would provide for more optimum usage. A second stage will maintain final effluent quality while the effluent from the first step gradually rises to near influent concentrations, utilizing all of the adsorptive capacity of the carbon. When the carbon in the first unit is fully utilized, it is then replaced with fresh carbon and put back on line as the final stage, allowing full utilization of the carbon in the other unit now in service as the first stage. The concept of staged adsorbers is of particular value when the designer is considering a water requiring treatment that contains a variety of contaminants exhibiting differing adsorption characteristics.

After the contact time has been established and the evaluation of the breakthrough curves has indicated whether a single bed or staged system is preferred, the designer can select the adsorber configuration.

If the breakthrough curve is steep, usually in the case of single or similar contaminants, the single fixed-bed downflow adsorber is the most economical choice. The contact time will establish the total carbon volume as noted above. By weighing considerations such as flow and carbon volume, the designer will select the vessel size and decide whether multiple units (operated in parallel) may be required.

The simplest downflow fixed bed is the gravity adsorber. Because the downflow unit operates as a filter and is not a pressure vessel, flow is limited to 2 gpm/ft$^2$. At 2 gpm/ft$^2$, the typical carbon depth is 4 ft, yielding a contact time of 15 min. The effluent from a gravity adsorber will require pumping if needed for a pressure water system. As the system depends upon gravity as the motive force, the gravity adsorber may require backwashing if suspended solids are present in the influent.

The fixed-bed system can also be contained in a pressure vessel. This vessel allows greater bed depth, which can be used for higher surface loading rates (up to 5 gpm/ft$^2$) while maintaining the same contact time or greater contact times. This system can also be operated at higher pressures, so the unit could be placed in line between the pump and downstream usage. The pressure drop through a typical granular carbon is provided in Figure 3–18.

If the breakthrough curve is gradual or discontinuous (has temporary plateau value) due to multiple contaminants, the designer may wish to specify a staged carbon adsorption system in order to obtain more optimum utilization of the adsorptive capacity of the carbon. The simplest staged adsorption system is two single fixed beds in series operation. If the mass transfer zone can be maintained within a single bed, then the second stage will be able to maintain effluent quality, while the carbon in the first stage is obtaining full use of its adsorptive capacity. When the carbon in the first stage is fully utilized, it is exchanged with fresh carbon and returned to service as the second stage. The fixed-bed downflow system has the added advantage of operating as a media filter with elimination of suspended solids in the effluent.

Another form of the staged bed system is the upflow moving-bed design. This system may be of use when long contact times are required and the breakthrough curve indicates that even a two-stage system is insufficient to provide economical use of the carbon.

In this process, the carbon is placed in a large coned-bottom vessel. The coned bottom is desirable because removal of carbon will be of the mass (plug) flow variety. A coned bottom will insure that the material at the side walls will move downward in the vessel at the same rate as that in the center. Flow from the center only, termed ratholing, results in uneven distribution of fresh and partially spent carbon in the bed and may cause premature breakthrough of the contaminant.

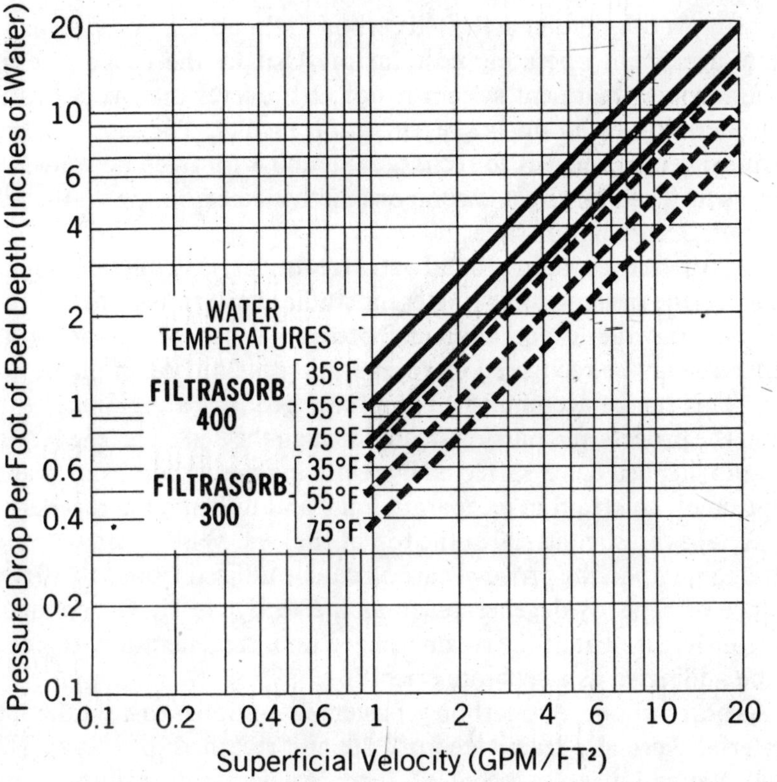


Figure 3–18. Downflow pressure drop through a backwashed segregated bed of Filtrasorb 300 and Filtrasorb 400. *(Courtesy of Calgon Carbon Corporation.)*

The water flow is directed upflow through the bed at rates up to 8 gpm/ft². At this rate, there may be a slight expansion of the bed. This will produce a small amount of carbon fines material in the effluent, which, if a problem, will require nominal filtration.

When a portion of the carbon is fully utilized, it is withdrawn from the bottom of the unit, and an equal volume of fresh carbon is placed at the top of the bed. This portion of the bed replaced may be anywhere from 5% to 50%, depending upon the breakthrough curve. Replacement volumes of 50% or greater usually indicate that a two-staged fixed-bed system may have been a better selection.

### Granular Activated-Carbon Replacement Considerations

The supply of granular activated carbon to the adsorption system may be a significant operating cost factor. Usually the usage rate in a groundwater treatment system is not at the level that would justify consideration of an on-site regeneration facility. The most common form of replacement is to recharge the unit with fresh activated carbon, which can be virgin carbon or reactivated carbon, with the reactivation occurring off site.

As is often the case in groundwater treatment, the contaminant may be a volatile organic solvent. Recent studies have shown that, if there are no nonvolatile contaminants present, much of the adsorptive capacity may be recovered by regenerating the carbon in place, or in situ. This can be accomplished in a fixed bed by withholding the unit from the process and passing steam through the bed. The regeneration in a moving bed can be affected by removing the carbon from the bed, exposing it to steam in a separate unit, and utilizing the regenerated material as fresh material at the top of the bed. With steam stripping, depending upon the process, carbon can be utilized from 5 to 10 times before its capacity degenerates to an ineffective level. The condensation and treatment of the steam utilized in the regeneration step needs to be addressed as a separate operation.

As noted above, one carbon replacement option is to have the spent material thermally reactivated off site and returned to service. Thermal reactivation is conducted in a furnace where temperatures of up to 1800°F are obtained. The advantage of thermal reactivation is that the organic contaminants are driven off the carbon and thermally destroyed. The regeneration facility should be equipped with a thermal afterburner to insure complete destruction of organic compounds and a scrubber to remove acid gases that may be present due to chlorinated solvents often found in groundwater.

The advantage of regeneration is the recovery of the granular activated carbon for further use. The regeneration step may be a custom operation, in which the spent carbon is regenerated as a batch and returned to the user. Pool regeneration involves collection and regeneration of many carbons to effect economy of scale, but the carbon returned is not the original material.

Virgin carbon should be used in applications involving treatment of the water for use (potable applications, for example), but reactivated carbon is generally less expensive in applications involving aquifer restoration, where the water is treated for discharge.

Regenerated carbon may also possess different adsorptive qualities than virgin-grade material, which may or may not be detrimental to the efficiency of contaminant removal. A proper evaluation of the regenerated material, followed by an overall economic comparison of the cost and effectiveness of the two processes, virgin carbon versus regenerated carbon, will determine the most economical approach.

### Operating Results—Case Studies

Granular activated carbon has been utilized successfully in many cases to treat contaminated groundwater. The case studies presented here show some of the wide variety of organic compounds and concentrations that may be present in groundwaters which can be removed effectively by granular carbon adsorption.

Table 3–6 summarizes actual cases in which carbon was used to treat a contaminated groundwater for drinking water use. All of these cases involved use of virgin-grade granular carbon to insure the purity of the water.

This review shows that in cases where one contaminant is present in substantial quantity, effective use of the carbon can be obtained with short contact times (less than 30 min) and a single fixed-bed system. This design provides for minimal equipment and carbon on-line, and subsequent minimum cost of treatment. In a vast majority of cases involving contaminant concentrations at levels of micrograms per liter, carbon usage rates are less than 0.5 lb/1000 gal.

If the carbon treatment is being utilized for drinking water purposes, chlorination is usually used after the adsorption process to insure against biological activity in the downstream distribution system. The operating costs for treating groundwater contaminated at levels of micrograms per liter range from $0.22 to $0.55/1000 gal. These operating costs reflect the amortization of the installed equipment and the replacement of the granular activated carbon as required.

Table 3–7 reviews cases of groundwater contamination in which contaminants are present in levels of milligrams per liter. These ex-

## TABLE 3-6. Carbon Adsorption with ppb Influent Levels.[a]

| SYSTEM NO. | CONTAMINANTS | TYPICAL INFLUENT CONC. (μg/liter) | TYPICAL EFFLUENT CONC. (μg/liter) | SURFACE LOADING (gpm/ft²) | TOTAL CONTACT TIME (min) | CARBON USAGE RATE (lb/1000 gal) | OPERATING MODE |
|---|---|---|---|---|---|---|---|
| 1 | 1,1,1-Trichloroethane | 143 | < 1 | 4.5 | 15 | 0.4 | Single fixed bed |
| | Trichloroethylene | 8.4 | < 1 | | | | |
| | Tetrachlorothylene | 26 | < 1 | | | | |
| 2 | Methyl T-butyl ether | 30 | < 5 | 5.7 | 12 | 0.62 | Two single fixed beds |
| | Di-isopropyl ether | 35 | < 1 | | | | |
| 3 | Chloroform | 400 | < 100 | 2.5 | 26 | 1.19 | four single fixed beds |
| | Trichloroethylene | 10 | < 1 | | | | |
| 4 | Trichloroethylene | 35 | < 1 | 3.3 | 21 | 0.21 | Three single fixed beds |
| | Tetrachloroethylene | 170 | < 1 | | | | |
| 5 | 1,1,1-Trichlorethane | 70 | < 1 | 4.5 | 30 | 0.45 | Two fixed beds in series |
| | 1,1-Dichloroethylene | 10 | < 1 | | | | |
| 6 | Trichlorethylene | 25 | < 1 | 2.0 | 35 | 0.32 | Single fixed bed |
| | *Cis*-1,2-dichloroethylene | 15 | < 1 | | | | |
| 7 | Trichlorethylene | 50 | < 1 | 1.6 | 42 | 0.38 | Two single fixed beds |
| 8 | *Cis*-1,2-dichloroethylene | 5 | < 1 | 1.91 | 70 | 0.25 | Two fixed beds in series |
| | Trichloroethylene | 5 | < 1 | | | | |
| | Tetrachloroethylene | 10 | < 1 | | | | |

[a]From O'Brien, Robert, and Ficher, J. L. There is an Answer to Groundwater Contamination. *Water/Engineering & Management*, May 1983.

**TABLE 3-7. Carbon Adsorption with ppm Influent Levels.[a]**

| SYSTEM NO. | | TYPICAL INFLUENT CONC. (mg/liter) | TYPICAL EFFLUENT CONC. (mg/liter) | SURFACE LOADING (gpm/ft²) | TOTAL CONTACT TIME (min) | CARBON USAGE RATE (lb./1000 gal) | OPERATING MODE |
|---|---|---|---|---|---|---|---|
| 1 | Phenol | 63 | < 1 | 1.0 | 201 | 5.8 | Three fixed beds in series |
| | Orthochlorophenol | 100 | < 1 | | | | |
| 2 | Chloroform | 3.4 | < 1 | 0.5 | 262 | 11.6 | Two fixed beds in series |
| | Carbon tetrachloride | 135 | < 1 | | | | |
| | Tetrachloroethylene | 3 | < 1 | | | | |
| | Tetrachloroethylene | 70 | < 1 | | | | |
| 3 | Chloroform | 0.8 | < 1 | 2.3 | 58 | 2.8 | Two fixed beds in series |
| | Carbon tetrachloride | 10.0 | < 1 | | | | |
| | Tetrachloroethylene | 15.0 | < 1 | | | | |
| 4 | Benzene | 0.4 | < 1 | 1.21 | 112 | 1.9 | Two fixed beds in series |
| | Tetrachloroethylene | 4.5 | < 1 | | | | |
| 5 | Chloroform | 1.4 | < 1 | 1.6 | 41 | 1.15 | Two fixed beds in series |
| | Carbon tetrachloride | 1.0 | < 1 | | | | |
| 6 | Trichloroethylene | 3.8 | < 1 | 2.4 | 36 | 1.54 | Two fixed beds in series |
| | Xylene | 0.2–0.5 | < 1 | | | | |
| | Isopropyl alcohol | 0.2 | < 10 | | | | |
| | Acetone | 0.1 | < 10 | | | | |
| 7 | Di-isopropyl methyl phosphonate | 1.25 | < 50 | 2.2 | 30 | 0.7 | Single fixed bed |
| | Dichloropentadiene | 0.45 | < 10 | | | | |

[a]From O'Brien, Robert, and Ficher, J. L. There Is an Answer to Groundwater Contamination. *Water/Engineering & Management*, May, 1983.

amples include cases in which chemical spills, landfills, and storage tanks have led to a more severe groundwater problem. The situations covered here have utilized reactivated-grade carbon, because the end use of the treated groundwater is not for drinking water but for discharge to surface water, recharge to the aquifer, or plant process use.

Due to the higher concentrations, and in most cases the presence of two or more predominant contaminants, the process of choice becomes the staged system, which insures more complete utilization of the activated carbon while maintaining the effluent at the desired level. The multistage system also allows for longer contact times required to meet the low effluent concentrations.

In most cases, pretreatment to the carbon system is not necessary. Filtration may be required if the water is high in suspended solids material and it is not desired to have the carbon bed utilized as a filter (i.e., due to higher backwash requirements). Also, pH adjustment may be required if the water has a high pH and contains mineral salts susceptible to precipitation in the carbon bed. These pretreatment needs would normally be determined in the evaluation procedures.

The operating costs for treating groundwater contaminated at levels of milligrams per liter range from \$0.48 to \$2.52/1000 gal. Again, these costs reflect amortization of installed equipment and replacement of the granular activated carbon as required. In most of these cases, the replacement of carbon will be regenerated material, which reflects some savings over utilizing virgin-grade material.

These results, showing removal of a wide range of organic compound contaminants to low or nondetectable levels, show that granular-carbon adsorption is a versatile groundwater treatment process.

## Application with Other Technologies

Carbon adsorption is a relatively expensive process. However, the inherent advantages of the technology make it particularly suited for low concentrations of nonvolatile components, high concentrations of nondegradable compounds, and short-term projects. When there is a variety of compounds, or when very low effluent levels are required, carbon adsorption can be combined with other treatment techniques.

Carbon adsorption may readily be combined with biological treatment to effect better overall performance. Powdered activated carbon may be added directly to the biological system, either to provide sites for organic compounds to adsorb in order to undergo biological degradation, or to remove refractory organic compounds that may be toxic to the system. In this case, the evaluation procedure is conducted considering specific organic compounds only. Finally, granular-carbon systems can be used to polish the effluent from biological systems to remove refractory compounds.

Carbon adsorption also serves as a complementary technology to air stripping. Granular activated-carbon systems can be utilized to treat air-stripping effluent water to remove volatile and nonvolatile components. Nonvolatile components such as phenols, pesticides, and other substituted aromatics that are not volatile can be removed in a carbon adsorption step. As air stripping is an equilibrium process, some concentrations of the volatile contaminant will remain following the treatment step, so carbon can be utilized to remove such contaminants to nondetectable levels. The utilization of air stripping as pretreatment to carbon adsorption enhances the life of the carbon. Because the more volatile contaminants are those less readily adsorbed, their removal allows for use of less carbon. In many cases, the application of both air stripping and granular activated carbon will be the most cost-effective solution.

Although this section has discussed treatment of contaminated waters, the application of granular activated carbon to treat gas streams is of importance as a complementary technology. Although this was discussed in the section on air stripping, further discussion is warranted here. Granular activated carbon has been proven to be effective in removing organic vapors from exhaust airstreams.

Granular activated carbons and systems for vapor adsorption are different from those normally used for liquid-phase systems. The carbon particles are usually larger to minimize pressure drop of the gas stream, and because contaminants are easily volatilized, the systems can be designed for in-situ regeneration. The evaluation of activated carbon for vapor-phase adsorption is similar, however, and isotherms for a variety of contaminants have been established. Pretreatment with condensers or dehumidifiers will also enhance the vapor-phase adsorption step by reducing water vapor content and reducing the volume of the gas to be treated.

## MISCELLANEOUS TREATMENT METHODS

Minor methods that can be used for treatment of organic contaminants in groundwater are mainly oxidation methods. A case can be made for the use of membrane systems or synthetic resins on groundwater. Both of these technologies are separation technologies, and they are normally applied when the compound removed has a high enough value to be recovered.

Ultrafiltration has been widely used for the separation of emulsions, mainly oil emulsions, from water. The filtrate will still contain all of the surfactants and therefore will require further treatment for groundwater. In industry, both the oil and the surfactants are reused.

Synthetic resins perform the same function as activated carbon. However, they can be made more selective, have a higher capacity, and are easier to regenerate. They are also much more expensive than carbon. When the compound to be removed has a high value, however, the extra cost is affordable.

Both of these technologies are available for groundwater projects, but their use will be very minor. Of more importance to the groundwater treatment field is the use of the various oxidation methods.

### Incineration

Incineration is the raising of the water and the contaminants to high temperatures in the presence of oxygen. At temperatures of between 1500 and 2000°F, essentially all organic compounds are oxidized. When the contaminated water contains approximately 20% organics, the solution will have a self-sustaining burn. At concentrations below this level, auxiliary fuel is required. Organic concentrations are rarely this high in groundwater. First, pure contaminant will be adsorbed in the unsaturated zone. Then the soluble compounds will be diluted by the groundwater. Nonsoluble compounds will stay out of the aquifer and be removed by pure-compound removal techniques.

Another problem is the high capital cost of incinerators. The groundwater site will have to be close to an existing incinerator for the application of small volumes of toxic materials to be possible. There are few portable incinerators available.

Incinerators can be combined with the other treatment technologies. Any of the technologies that separate and concentrate the contaminant can be used as a precursor to incineration. Also, one design (6) used incineration for treatment of the airstream off a stripper. With the water removed from the contaminants, the amount of auxiliary fuel was kept to a minimum. However, the economics presented by the paper make incineration a doubtful choice for treatment of groundwater.

One area that has been using incineration is the destruction of toxic chemicals in soil. This treatment is limited to cases where only the top layers of soil have been contaminated. Also, the contaminants are not soluble in water and therefore have remained at the surface, where they were originally released. In both cases, a limited amount of contaminated soil has been created. Incineration is an expensive process, and unique situations must exist in order for it to be used. On-site incineration can be used when the engineer decides that the contaminated soil is too dangerous to transport over the roads, or when the nearest landfill is too far away.

### Chemical Oxidation

Three chemical oxidants have been widely used for industrial treatment: chlorine, hydrogen peroxide, and ozone. One of the main contaminants being treated in groundwater is chlorinated hydrocarbons in the low parts-per-billion range. Chlorine oxidation produces chlorinated hydrocarbons that would be unacceptable in groundwater. Therefore, chlorine will find very limited use in groundwater treatment.

Hydrogen peroxide is readily available and works very well on organic compounds with double and triple bonds. Ring compounds (phenol and benzene) and cyanide are two examples of compounds that can be oxidized by peroxide. The costs and handling of peroxide will limit its use to small flows and short-term projects.

Ozone is the strongest of the oxidizing agents. With sufficient time, ozone can eliminate any organic compound. However, because of high capital and operating costs, ozone will also find limited direct use in groundwater cleanup.

One technology advancement that has recently been applied to groundwater is the use of UV-ozone. A UV-ozone reactor places UV lamps in the water at 3- to 6-in. intervals. Ozone is added to the reactor along the bottom. The UV light catalyzes the ozone reaction. This technology has had a tremendous amount of research but limited use on industrial waste.

However, for groundwater this technology may have more value. The reason for this is that the rate of reaction for oxidizing low concentrations of chlorinated organics is extremely fast. UV-ozone reactors have been reported to reduce 100 ppb of trichloroethlene (TCE) to 0.6 ppb with less than one-minute residence times. There is only one full-scale installation of this technology on groundwater at the present time. However, UV-ozone technology for chlorinated solvents and pesticides looks very promising.

Other technologies will develop as the groundwater market matures. These will take the form of improvements to existing technology and the production of completely new technology. The design engineer must keep abreast of these developments in order to provide the most cost-effective treatment.

## REFERENCES


1. Davis, J. B., et al. The Migration of Petroleum Products in Soil and Groundwater. American Petroleum Institute, December 1972.
2. Yaniga, Paul M. Groundwater Abatement Techniques for Removal of Refined Hydrocarbons. *Hazardous Wastes and Environmental Emergencies Proceedings*, March 1984, HMCRI.
3. Verschueren, Karel. "Handbook of Environmental Data on Organic Chemicals." New York: Van Nostrand Reinhold, 1983.
4. Anonymous. Underground Spill Cleanup Manual. American Petroleum Institute, API Publication 1628, June 1980.

5. Uhler, R. E., et al. Treatment Alternatives for Groundwater Contamination. James M. Montgomery, Consulting Engineers.
6. Nicholson, B. C., et al. Henry's Law Constants for the Trihalomethanes: Effects of Water Composition and Temperature. *Environmental Science and Technology,* 18:7–518 (1984).
7. McGary, F. J., and Lamarre, B. L. Groundwater Reclamation, Gilson Road Hazardous Waste Disposal Site, Nashua, New Hampshire. Water Pollution Control Federation Conference, Atlanta, 1983.
8. Stenzel, Mark. Letter of correspondence to Evan Nyer, August 22, 1984.

# 4
# Treatment for Organic Contaminants: Biological Methods

Pure-compound recovery is the least costly method of treatment for organic contaminants, but once the contaminant has entered the aquifer, biological treatment is the lowest-cost method of treatment per pound of organic removed. This cost advantage has made biological treatment the most widely used method of treatment for industrial and municipal wastewaters.

However, there are restrictions on when a biological method may be used. The main restriction with biological treatment is that it must be run 24 hr per day, 7 days per week. A biological treatment system cannot be turned on and off, because it is a living system. The reactor must have a constant source of food (the organic contaminant) and nutrients. Another restriction is that the bacteria must be grown to a sufficient concentration in order to effectively remove the contaminants. This startup period can take two to eight weeks, an unacceptable delay on short-term projects. Also, the fact that standard biological reactors are not designed for influent concentrations below 50–75 mg/liter will again limit the application of biological technology.

The existing biological technology will still have application to groundwater cleanup. In addition, new developments in the area should spread the use of biological treatment. None of the above restrictions apply to in-situ use of bacteria. (In-situ treatment will be covered in the second section of this chapter.) Also, work on low-concentration biological reactors has already extended the use of biological methods.

Even with the restrictions, the biological treatment system will destroy organic compounds (the compounds are not transferred to another medium) at about one-twentieth the cost of carbon adsorption. Biological treatment is also less costly than air stripping when the gas stream has to be treated.

## BIOLOGICAL TREATMENT

### Bacteria

Bacteria can be grown in two major environments, aerobic and anaerobic. The difference between the aerobic and anaerobic bacteria is that the aerobic bacteria use molecular oxygen as their terminal electron acceptor. Anaerobic bacteria use some other compound as their terminal electron acceptor. One common anaerobic terminal electron acceptor is carbon dioxide, which is converted to methane in the biological reaction.

Anaerobic bacteria are best suited for degrading organics in the 4000-mg/liter to 50,000-mg/liter range. Above that level, some form of evaporation or incineration should be employed. Anaerobic systems will find limited use in groundwater contaminations, although landfill leachates will be one exception. Also, there has been some work on anaerobic bacteria degrading chlorinated hydrocarbons in soils (1). The in-situ section will discuss the role of aerobic, anaerobic, and facultative (able to grow in both environments) microorganisms in soils.

Aerobic bacteria are classically used for organic concentrations between 50 and 4000 mg/liter. With small waste flows, the concentration can be extended to 10,000 or even 15,000 mg/liter. The small size of the treatment systems and the life-cycle concentrations of the organic will extend the concentrations in which aerobic bacteria are applied in groundwater treatment systems. Also, ongoing work by the author and others shows the possibility of extending the use of aerobic bacteria to influent concentrations as low as 10 mg/liter (1,2). This work needs further pilot tests and a full-scale demonstration.

No matter how the microorganisms are used, the basics of biological treatment remain the same. Figure 4-1 summarizes the biological process. Organics are put into contact with bacteria. The bacteria use the organics as a source of energy and building blocks for new bacterial cells. Macronutrients, nitrogen and prosphorous, and micronutrients are also required by the bacteria. The bacteria produce carbon dioxide, water, and new bacteria from the organics that they consume. The environment in which the bacteria grow must also meet the following criteria: pH = 6–9; temperature = 45–105°F; all toxic organics below toxicity level; and all toxic metals below toxicity level.

Bacteria are able to change their environment. For example, 500 mg/liter of phenol is normally toxic to bacteria, but this does not mean that a groundwater containing over 500 mg/liter of phenol could not be degraded by bacteria. If the bacteria are grown in a completely mixed tank, the concentration of phenol in the reactor is the effluent concentration, not the influent concentration. The bacteria will be in contact with less than 1 mg/liter of phenol. The bacteria will degrade the phenol at such a rate that it never reaches the toxic concentration. Of course, if the toxic organic is nondegradable, the bacteria will not be able to change their environment. The compound will reach a toxic concentration and the bacteria will be affected.

Biological systems can also affect the concentration of toxic, heavy metals. Toxic metals must be in the soluble form in order to affect

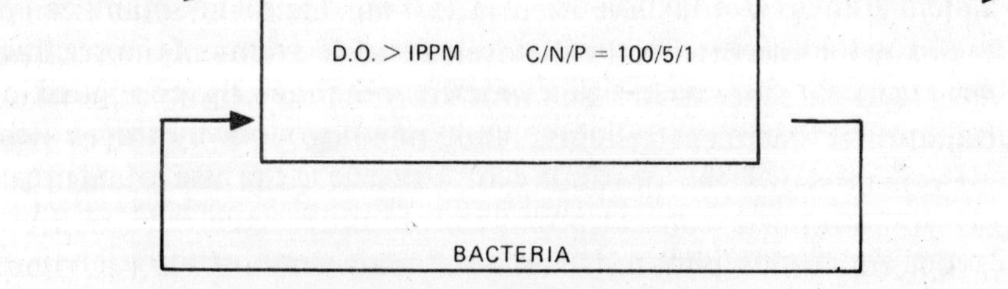


Figure 4-1. Aerobic biological treatment.

bacteria, but most heavy metals are not soluble at the pH range in which bacteria grow. Once the metal enters the neutral pH range of the biological system (if the influent water is too acidic, then basic compounds must be added to the system in order to maintain the proper pH for the bacteria), natural water chemistry takes over and the metals precipitate. Metals can be maintained in solution under neutral pHs with the help of a chelate. Most chelates are organic, so the bacteria usually can degrade these compounds and remove them from the water. Once again, the water chemistry can take over and remove the metal. Bacteria also have the ability of storing the metal in their cells without harm. This method may also be employed to maintain the environment at below-toxic levels.

Bacteria also have a tendency to neutralize the water in which they are growing. In addition, engineering methods can be employed to cool or retain heat in the reactor to maintain the proper temperature. When any of these methods is used, the bacteria and the system must not be overloaded. If the particular problem is too much for the natural system, some type of pretreatment must be employed to make sure that the bacteria are growing in the proper environment.

## Biological Reactors

Biological reactors can be separated into two main types: suspended-growth reactors and fixed-film reactors. With suspended-growth reactors, the bacteria are grown in the water and are intimately mixed with the organics in the water. Fixed-film systems grow bacteria on an inert support medium, and the water with the organics passes over the film of bacteria. Both have advantages and disadvantages.

The easiest unit operation to set up for a groundwater treatment system is an aerated lagoon or basin. An existing pond or tank can be used for the reactor. In some cases, portable swimming pools have been used for the aeration tank. Figure 4–2 shows the configuration for an aerated lagoon. The contaminated groundwater enters the aerated vessel. Bacteria in the reactor degrade the organics and create new bacteria. The liquid residence time in the reactor, which is equal to the bacterial residence time, must be sufficient for the bacteria to reproduce before they exit with the water. The longer the residence time, the lower the residual organic concentration will be. Figure 4–3

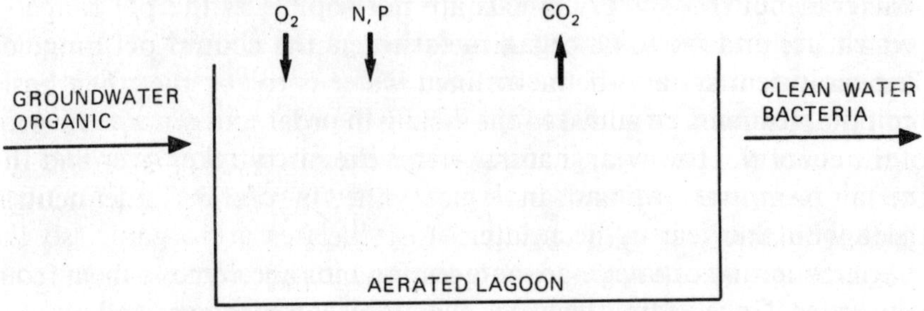


Figure 4–2. Aerated lagoon.

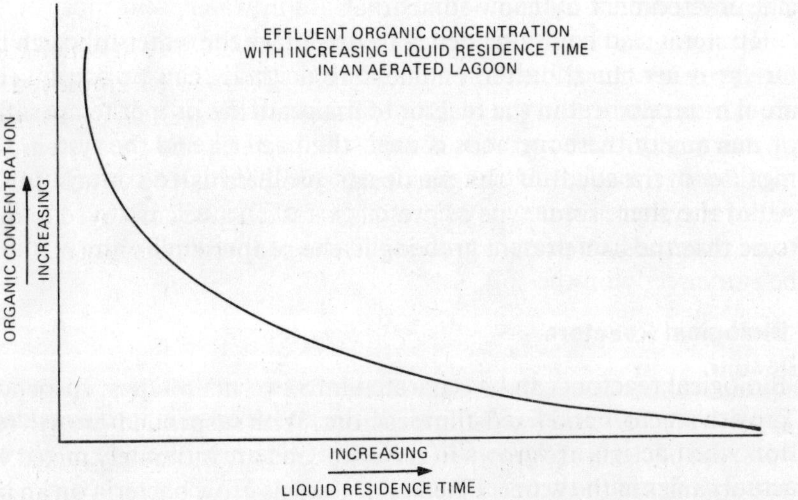


Figure 4-3. Effluent organic concentration with increasing liquid residence time in an aerated lagoon.

shows the relationship between effluent organic concentration and residence time.

Oxygen is supplied to the tank with a surface aerator or air diffusers. Sufficient power must be supplied to provide an adequate oxygen concentration, 2 mg/liter, and/or to keep the tank completely mixed. With short-residence-time reactors, oxygen supply is usually the controlling factor. Mixing is usually the controlling factor for long-residence-time reactors. Two days is the shortest residence time that should be used. Longer residence times will only be limited by total flow and the cost of power to mix the tank. The main operating cost (other than personnel) for any aerobic biological treatment system is the power requirements for oxygen supply or mixing.

Two problems exist with the use of an aerated-lagoon design. First, the degree of treatment is limited by the limited bacterial residence time. Lagoons can be expected to remove only 50–70% of the biodegradable organics. Second, the bacteria that are created in the reaction will be in the water when it leaves the reactor. A clarifier can be added to the system to remove the solids, but bacteria grown in an aerated lagoon do not settle readily. These limitations make the lagoon design suited for use only in in-situ treatment situations. Even in in-situ treatment, a clarifier is added to limit the amount of solids that are returned to the soil. Direct discharge and even discharge to another treatment system require more thorough treatment.

These problems can be solved by separating the liquid residence time from the bacterial residence time. Figure 2–2 showed that if a clarifier is added to remove the bacteria from the water stream and they are then returned to the aerated reactor, the bacterial residence time is made separate. From Figure 2–2, the liquid residence time ($R_L$) is

$$R_L = \frac{V}{Q} \tag{4–1}$$

The bacterial residence time ($R_B$) is

$$R_B = \frac{X \bullet V}{Q_W \bullet X_R + (Q - Q_W)X_E} \tag{4–2}$$

The bacterial residence time is controlled by the loss of bacteria due to wasting the settled bacteria from the clarifier and by the uncontrolled loss of bacteria over the top of the clarifier. When bacteria are returned to the aeration tank, the process is called *activated sludge* (Figure 4–4).

Activated sludge is the most widely used method of biological treatment in the wastewater treatment area. The basic advantages are that the process produces low effluent concentrations, the system can treat

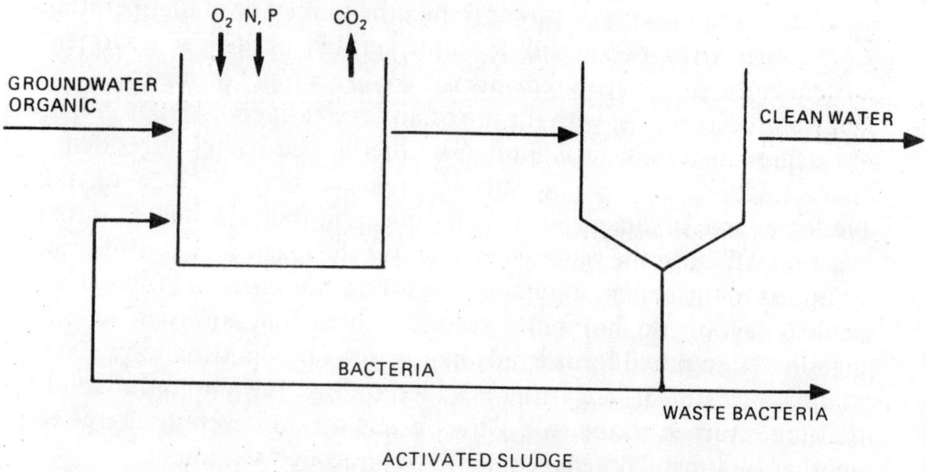


Figure 4-4. Activated sludge.

many organics at the same time, and the same equipment can be used for a variety of influent conditions. The main disadvantages are the cost of manpower to keep the system adjusted to the influent conditions, the relative cost of oxygen transfer compared to fixed-film systems, and the critical need to keep the bacteria in a growth stage in which their settling characteristics are at a maximum.

Chapter 2 presented a design example of an activated-sludge system over the life cycle of a groundwater cleanup. The activated-sludge process can remove 85–90% of the biodegradable organics from the influent, and 99+% of specific organic compounds. Effluent concentrations of 10–30 mg/liter of biochemical oxygen demand (BOD), a general measurement of biodegradable organics in water, can be expected with a well-run system. Effluent concentrations of specific compounds will depend on the particular compound. For example, effluent phenol concentrations from an activated-sludge system can be as low as 0.02 mg/liter.

Certain compounds, for example sugars and alcohols, degrade very quickly in a biological system. Other compounds require longer contact times with the bacteria in order to degrade. The easier a compound can be assimilated by the bacteria, the faster and more efficiently the bacteria can turn that compound into new bacteria. When compounds are hard to degrade, (refractory), the bacteria need more residence time in order to replace their lost numbers. Another way to look at this is that the bacteria must first remove the easily degradable organics before they are willing to produce the enzymes necessary to degrade the refractory compounds. They must be a little hungry before they go after these organics. In the design of the treatment plant, this can be represented by the food-to-microorganism ratio, *F/M*. From Figure 2-2, the formula would be

$$\frac{F}{M} = \frac{Q \bullet S}{V \bullet X} \tag{4-3}$$

In activated sludge, all of the compounds are being degraded at the same time in the completely mixed tank.

The reader can use either of these models, bacterial residence time (sludge age) or food-to-microorganism ratio, to help understand how to accomplish this concurrent degradation. One of the main advantages of the activated-sludge process is that the bacterial residence time and the *F/M* can be controlled to accommodate the degradation of a variety of compounds even if they have different degradation rates.

The life-cycle design example in Chapter 2 showed that this versatility can be maintained over a variety of influent conditions. With innovative engineering, the process can be made to work over a large percentage of the life cycle of the project.

The main problem with the activated-sludge design is the critical need to keep the bacteria in a form in which they settle readily. If the bacteria do not settle properly, the clarifier will not be able to remove them from the water stream. If the bacteria are not separated from the water stream and returned to the aeration basin, the whole process fails. This does not mean that the bacteria have lost their ability to degrade organics. The bacteria will still be able to degrade the compounds. But if the bacterial residence time cannot be separated from the liquid residence time, the end result is just an aerated lagoon and the corresponding removal rates.

In order to maintain the settling properties, two things are necessary: The environment in which the bacteria grow must not have any major changes, and the bacteria must be grown at the proper sludge age that promotes flocculation. In groundwater treatment, the influent has very little variation on a day-to-day basis. There is normally no need for equalization as is used in wastewater treatment. The main problem using activated sludge with groundwater is the life-cycle concentration and growth of the bacteria in their flocculant stage during the entire project. The example in Chapter 2 showed one way to overcome this problem. It is important for the design engineer to design the activated-sludge treatment system to be operable over as much of the life cycle as possible. Even with a good design, the activated-sludge system will still require a relatively high level of operator attention to insure that the system is operating in the correct manner.

Another way to use bacteria in the cleanup is to set up a fixed-film system. In fixed-film systems, large-surface-area inert support media are placed in the reactor. Bacteria naturally attach to and grow on any surface provided to them. The water, with the contaminants, once again passes through the biological reactor, and clean water comes out. There are two important advantages of the fixed-film systems: The bacteria are maintained in a high concentration without the need of a clarifier, and oxygen can be supplied at lower costs. This is accomplished by trickling the water over the media in a tower design or passing the bacteria through the water and back to the air in a rotating biological contractor (RBC). The oxygen, in both cases, naturally transfers into the thin film of water formed.

A third advantage is the general ease of operation. A fixed-film system does not require the operator attention that an activated-sludge system does. Bacteria will grow attached to the media and remove organics from the water over a wide range of operating conditions. When there are too many bacteria in the system, the bacteria will sluff off of the media and leave the reactor with the water (a clarifier can be used to remove the solids before final discharge).

There are several technical disadvantages with the fixed-film reactors. Fixed-film reactors are plug flow reactors. The water comes in at one end, passes by the bacterial film, and exits the other end of the reactor. The influent end of the reactor sees the high influent concentration of the contaminant. In completely mixed reactors, the influent is immediately mixed with the contents of the tank. The influent concentration may be toxic, or pockets of high-concentration material may be found as the groundwater is brought up from the ground. The bacteria in the fixed-film reactor will be subjected to the full concentration. Recycling of the effluent water can be used to minimize this effect, but it also adds to the cost of operation.

Another problem with fixed-film systems is that they will not remove as high a percentage of the influent contaminant as an activated-sludge system. Specific chemical removal is very important in groundwater treatment. General removal of organics will be important, depending upon the final disposal of the water. The design engineer can expect 70–85% BOD removal and 85–95% removal of a specific organic. The lower the influent concentration, the less percent removal can be expected.

There are also practical disadvantages to using a fixed-film system on groundwater treatment systems. First, the design engineer will not be able to use existing tanks or portable tanks as part of the system. The fixed-film system will have to be built on site and then disassembled at the end of the project. As we discussed in Chapter 1, resale or reuse of equipment is important in groundwater systems.

As the groundwater market expands, advances in technology specifically suited for groundwater treatment can be expected. One example of biological technology suited for groundwater is the submerged fixed-film treatment system. Figure 4–5 shows one system. The bacteria are maintained in the reactor by growing on plastic media. The media are submerged under the water in the aeration tank. Air lift pumps circulate the water below the media to the top of the media, providing complete mixing and oxygen to the system. This system has the advantages of a fixed-film system and an activated-sludge system without the disadvantages of either. It has no moving parts, it is portable, and it is self-regulating.

Advances in low-concentration biological reactors will probably be in the fixed-film area of design, and other technological advances can be expected in this field as well. It is important for the design engineer to keep abreast of these advances in order to supply the best equipment to the treatment system.

### Applying Biological Treatment

Biological treatment should be applied when possible. Biological treatment is an inexpensive method when designed correctly. It is a

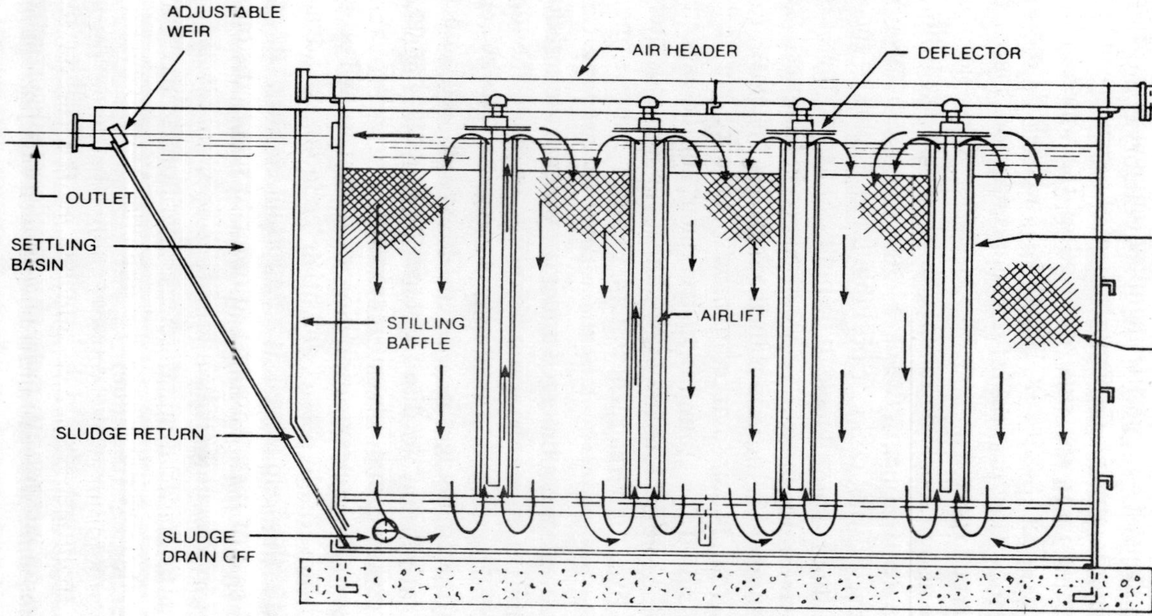


Figure 4-5. Fixed Activated-Sludge Treatment.[8] *(Courtesy of Smith & Loveless, Inc.)*

destruction method. The organics do not exist after being removed by the system. Many organics can be treated at the same time. The bacteria that are left behind will continue to clean up the soil and the aquifer after the treatment system is shut down. It is the natural system. Nature uses bacteria to break down complicated organic compounds (cellulose from trees and plants is a refractory compound) in order to recycle the components.

However, there are limitations to the process, and the design engineer should make sure that biological treatment is applicable to the particular situation. Biological treatment is not applicable to all organic compounds. It should not be applied when the treated water is to be used for final consumption by humans or animals unless the water is carefully treated afterward to remove all of the bacteria. Biological systems require two to six weeks to start up. Bacteria are difficult to apply to short-term projects other than in-situ treatment.

There are several steps that the design engineer should take before deciding on biological treatment for the cleanup of a particular organic compound:

1. Search literature for the degradability of the compound
2. Run general organic concentration tests—BOD, COD, TOC
3. Run treatability studies
4. Select biological process to be applied
5. Set up on-site biological treatment system

The design engineer must first decide if bacteria can degrade the compounds that have been found in the groundwater. This can be accomplished by any of the first three steps listed above. Probably, all three steps should be used, except in cases involving easily degradable compounds such as sugars and alcohols.

A considerable amount of work has been performed on the degradability of different organic compounds. The problem is that, on the refractory compounds, the literature does not always agree. Table 4–1 lists a compilation of several literature sources on the degradability of 20 specific organic compounds. As can be seen from Table 4–1, the more substitutions (halogen or other inorganic group replacing the hydrogen molecule attached to the carbon) on an organic, the more difficult the organic is to degrade. As long as one

**TABLE 4-1. Biodegradability of Specific Organic Compounds.**

| COMPOUND | BIODEGRADABILITY |
|---|---|
| 1. Acetone | degradable[a] |
| 2. Benzene | degradable[a] |
| 3. Carbon tetrachloride | nondegradable[a], refractory[b] |
| 4. Chloroform | nondegradable[c], degradable[b] |
| 5. Methylene chloride | degradable[b] |
| 6. Chloro benzene | degradable[a] |
| 7. Ethyl benzene | degradable[a] |
| 8. Hexachloro benzene | nondegradable[a] |
| 9. Ethylene chloride | refractory[a] |
| 10. 1,1,1-Trichloroethane | nondegradable[b], refractory[c] |
| 11. 1,1,2-Trichloroethane | refractory[c] |
| 12. Trichloroethylene | refractory[c] |
| 13. Tetrachloroethylene | nondegradable[b], refractory[c] |
| 14. Phenol | degradable[a] |
| 15. 2-Chlorophenol | degradable[a] |
| 16. Pentachlorophenol | refractory[a], degradable[b] |
| 17. Toluene | degradable[a] |
| 18. Methyl ethyl ketone | degradable[a] |
| 19. Naphthalene | degradable[a] |
| 20. Vinyl chloride | refractory[d] |

[a] From Verschueren, Karel. "Handbook of Environmental Data on Organic Chemicals." New York: Van Nostrand Reinhold, 1983.
[b] From Kincannon, D. F., et al. Predicting Treatability of Multiple Organic Priority Pollutant Wastewaters from Single Pollutant Treatability Studies. 37th Purdue Industrial Waste Conference, May 1982.
[c] From Tabak, H. H., et al. Biodegradability Studies with Organic Priority Pollutant Compounds. *Journal Water Pollution Control Federation,* 53:10–1503 (1981).
[d] From Krupka, M. J. Report to the Office of Technology Assessment Congress of the United States. Correspondence from Cytox Corporation, April 10, 1981.

literature source finds the compound degradable, the design engineer should proceed with the evaluation. Even if no previous successes can be found, the engineer may want to apply certain degradation-enhancement techniques in the treatability studies to promote degradation of the compound.

Before the treatability study, a general organic analysis should be performed. The biochemical oxygen demand (BOD), a chemical oxygen demand (COD), and a total organic carbon (TOC) tests should be run. These tests will tell the engineer the total amount of organic in the groundwater, not just the specific compound. The design of the treatment system is based on the total amount of organic in the water. The ratio of the results from these tests will also give the design engineer some idea about the degradability of the organic compound in the groundwater. The ratio between the test results should be about: 1:2–3:1; for BOD:COD:TOC. Low values for BOD may indicate nondegradable organic material present. Low values for the TOC may indicate inorganic oxygen demand present.

Treatability studies should almost always be run before a full-scale treatment system is set up. These studies will not be directly scalable to the full-scale system, but they will give an accurate picture on the ability of the bacteria to degrade the various organics in the groundwater. The studies will also show if there is something unexpected in the water that will be toxic to the bacteria.

Treatability studies can also be used to try different sources of bacteria on the groundwater. Bacteria can be obtained from various wastewater treatment plants in the geographic area of the cleanup. A wastewater treatment plant that is already degrading the organic contaminants found in the groundwater is the preferred source of bacteria. A plant treating the same family of compounds would also be a good seed source, for example, the groundwater has phenol in it and a local biological treatment plant treats benzene. A soil sample from the surface where the original spill occurred is also a good source of bacteria.

When an unusual compound must be degraded, samples of bacteria can be shipped in limited quantities from treatment plants and other spill cleanups across the country that have treated similar compounds. There are also commercial bacterial products that can be purchased from various companies. These products are available off the shelf. While there are a variety of products available that can degrade a wide range of organic compounds, the products do not degrade any compounds that natural bacteria can not degrade. The main advantage of the commercial bacteria is convenience. A high concentration of stable bacteria can be purchased and shipped to the project site to be used when needed. When the cost of bacteria is compared to the total cost of the project, the cost of commercial bacteria is about 5% of the total cost. This cost can be even less if the bacteria are only used to start up the biological reactors.

One final technique that can be tested in the treatability study is the addition of a cometabolite. Sometimes the bacteria will not recognize the organic compound in the groundwater as food. Since they do not recognize that there is a food source available, they will not turn on their enzyme systems that can break up the compound and assimilate it as food. A compound that the bacteria do recognize as food and that is similar in structure to the original compound can be added to the treatment system. The enzymes produced for the added compound may also work on the original compound. For example, suppose the groundwater contains trichlorophenol and the bacteria do not degrade it. Phenol is added to the sample, and the enzymes that the bacteria produce for the phenol are also successful in breaking the ring structure of the trichlorophenol. The cometabolite allows the biological system to degrade the compound.

Treatability studies can be done in a variety of hardware. Anything from shake flasks to respirometers can be used to determine the degradability of the compounds under various situations. A preferred method of the author is the "bag" reactor, Figure 4-6. A felt cloth bag is placed inside of a bucket or barrel, and influent to the reactor

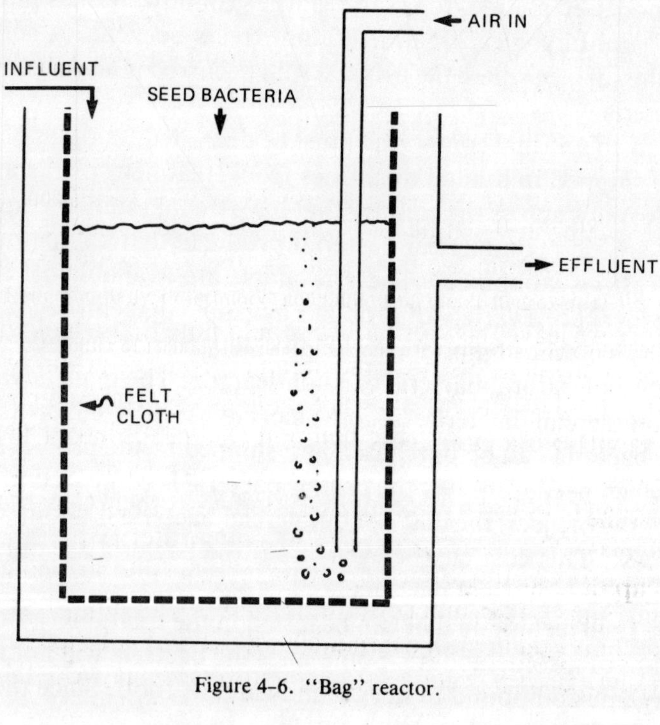


Figure 4-6. "Bag" reactor.

goes directly inside the bag. The water travels through the felt and out through the side of the bucket. The water level in the reactor is controlled by the height of the effluent line in the bucket.

Oxygen and mixing are provided by placing an air source at the bottom of the bag. Seed sources of bacteria can easily be placed in the bag. The felt cloth retains the bacteria inside the reactor, and bacteria grow on the felt and in the reactor itself. As can be seen, the reactor is easy to run and can be set up anywhere. The felt does, however, become clogged over long periods of time, so the reactor is not a good design for a full-scale plant.

When possible, a pilot plant should be run on the groundwater to confirm the design of the treatment system. The problem is that many groundwater cleanups are relatively small flows. Chapter 6 gives an example of overcoming this problem by using part of the full-scale system as the pilot plant. This way, the pilot plant confirmed the applicability of biological treatment. At the same time, all of the parts in the pilot plant were used on the full-scale plant, minimizing the costs for the total project. The biological treatment example in Chapter 6 covers this design example in detail.

Biological treatment will often be used in conjunction with other treatment methods. For strict control of specific organics, it is advisable to use biological treatment followed by carbon adsorption or other technology. The biological system removes most of the organic compounds at lower cost, and the carbon system insures that no compounds escape the treatment system. In these cases, the exact effluent concentrations from the biological treatment system are not critical and a pilot plant may not be necessary.

Before we leave this section, we should note that a biological treatment system will also strip some organics from the groundwater. The air-water contact for oxygen transfer will also provide air-water contact for stripping of organic compounds with relatively low vapor pressures.

It has been the experience of the author and others (3,4) that the rate of degradation is faster than the stripping rate. If the bacteria can degrade a compound, then the driving force behind stripping the compound is the equilibrium concentration of the organic in the reactor. Compounds that are not degradable will have a high equilibrium concentration. If they have a low vapor pressure as well, they will be stripped from the reactor. Compounds that are degradable will have a low equilibrium concentration, and the driving force for stripping will be minimal.

The design engineer should be careful to distinguish between the two removal mechanisms in the design of the treatment system. In situations where there are no regulations covering the discharge of organics to the air, this dual function of the biological system can be an advantage. However, the necessary design additions must be used when the air discharges are regulated. These were covered in Chapter 3, under Air Stripping.

## IN-SITU TREATMENT

In-situ treatment is the destruction, neutralization, and, in general, rendering harmless of the contaminants while not moving them. The contaminants are treated in place. Theoretically, in-situ treatment can be applied to inorganic and organic contaminants. However, in-situ treatment is practically limited to biological destruction of organics. Therefore, in-situ treatment will be covered in this chapter even though treatment of inorganics will be discussed.

When the contamination is high pH, low pH, or heavy metals, the main treatment method is to first change the pH (see Chapter 5). In theory, the acid or base could be placed below ground in the path of the groundwater. The groundwater would change its pH as it came in contact with the neutralizing agent. However, groundwater flows through an aquifer in a plug flow manner. There is very little mixing in an aquifer. The neutralizing agent would be pushed ahead of the groundwater flow and only come into contact with the leading edge of the plume. In reality, the groundwater must be brought to the surface and neutralized above ground. The treated water can then be returned to the aquifer to force the plume back to the central well. Treatment takes place above ground, not in-situ.

Another theoretical method would be to place an immobilized neutralizing agent underground. A slurry wall, made up of limestone, could be placed in the path of the contamination plume. This may work for a simple adjustment of an acid plume. The cost, however, would be very high. Also, in the case where heavy metals were present in the water, the precipitation of the metals would clog the slurry wall.

In addition, not many regulators would be satisfied with leaving the heavy metals below ground.

The one case where inorganic material can be treated in situ is in the unsaturated zone. The ground can be economically treated in place to a depth where heavy equipment can blend it. The neutralizing agent is placed on the contamination site, and the heavy equipment, normal farming equipment, is used to mix the agent into the soil. But once again, heavy metals are a problem.

Chemical treatment of organic material has the same problems when applied as an in-situ technique. Contact and mixing limit the effectiveness of the reaction. The only time in-situ techniques provide an added advantage to a treatment technology is with the application of biological treatment. In-situ biological treatment has been well documented (5). The following are the main advantages of in-situ biological treatment:

1. Cost effectiveness
2. Minimal disturbance to existing site
3. On-site destruction of contaminants
4. Continued treatment after shutdown of the project
5. Permanent solution

### Biological In-Situ Treatment

When applying in-situ biological treatment, it is important for the design engineer not to limit his thinking to below ground. Application of in-situ techniques require aboveground operations as well as below-ground operations. All of the bacterial requirements discussed in the biological treatment section of this chapter have to be applied to in-situ treatment. The design engineer must satisfy four main requirements:

1. Bacteria
2. Oxygen
3. Nutrients
4. Environment

Bacteria do all of the work in biological treatment. The design engineer must make sure that there are sufficient bacteria to consume all of the organic contaminants in a timely manner. The longer the project takes, the higher the cost of the cleanup. The design engineer will have the choice of enhancing the growth of the existing soil bacteria, growing large amounts of natural or imported bacteria in aboveground reactors, or applying commercially available bacteria.

The bacteria must be placed in contact with the organics, because bacteria are not highly mobile. The bacteria will eat the organics in their immediate vicinity and produce water, carbon dioxide, and new bacteria. But once the food is gone, the bacteria will shut down. They cannot pack up and move to the next source of food. The design engineer must get the bacteria to all of the organics in an in-situ cleanup.

Excess bacteria will also be produced, and when large amounts of organics are present, the excess bacteria must be removed. Wells and aquifers are made up of small openings. To keep these openings free for water flow, a limited amount of bacteria must be allowed to fill these open spaces.

The bacteria will require oxygen. Aerobic bacteria will degrade organic compounds at a faster rate and leave a lower concentration of specific organics. In addition, aerobic bacteria will yield more new bacteria per pound of organic consumed. The faster bacteria are produced, the lower the amount of time required for startup. Some work recently has shown that anaerobic bacteria may be able to treat chlorinated hydrocarbons better than aerobic bacteria (1). However, in most cases, in-situ treatment will be aerobic.

The bacteria will require nutrients. The design engineer does not want the level of nutrients to limit the rate of growth of the bacteria. Biological treatment should always be designed to have the organics as the growth-limiting factor. Most aquifer contaminations will be produced by the release of pure organic material. It is very unlikely that sufficient nutrients will be available for the bacteria in the spilled material, or in the unsaturated zone and the aquifer. Nutrients will have to be provided for the bacteria, and the design engineer must get the bacteria and the nutrients to the organics.

All of the environmental conditions must be appropriate for bacterial growth. Once again, the environmental conditions should not limit the growth rate of the bacteria. The organic should always limit growth. In-situ treatment requires that there be no toxic conditions present in the soil or in the aquifer. The pH should be in the correct range, and no toxic organics or inorganics should be present. Temperature must also be maintained in the correct range. This will limit the treatment of surface soils during the winter in certain regions of the country. However, groundwater is insulated from the swings in the air temperature, and biological activity can be maintained throughout the year in the aquifer.

Soil conditions provide advantages for biological growth. An advantage of in-situ treatment in the unsaturated zone is that the bacteria can tolerate much higher concentrations of toxic organics and inorganics. This is due to the low water content of the soil. Bacteria are affected only by compounds that are in water, and most of the compounds in the unsaturated zone are attached to the soil particles. On the other hand, bacteria are able to degrade the compounds that are not directly in the water. Land farming of oils is a prime example of this ability. Although the concentration of phenol may be 10,000 mg/liter in the soil, the bacteria do not see that high a concentration. All of the phenol is not in the water part of the soil. However, when toxic materials are present in the spill area, a treatability study should be run directly on the soil.

## Design

The design engineer must take all of the above criteria into consideration when designing an in-situ treatment cleanup. We will divide the actual design into three sections:

1. The aquifer
2. The unsaturated zone
3. Aboveground equipment

**The Aquifer.** The design engineer must control the water movement in the aquifer to insure the control of the contamination plume and to supply a source of oxygen, nutrients, and bacteria. In Chapter 1 we discussed the plume of contamination and the ways to control the movement of the plume by drawing water from central wells and placing the treated water in wells at the outside of the plume. For in-situ treatment to work, not only must we control the movement of the plume, we must also get bacteria, oxygen, and nutrients spread throughout the affected area of the aquifer.

The first design paramater to determine is the flow out of the wells. In aboveground treatment, this would be the flow to the treatment system. There are two main differences between flow for an aboveground treatment system and in-situ treatment. First, the flow through the aquifer should be as fast as possible. In-situ treatment takes anywhere from between two and six recirculations of the water present in the aquifer to clean the aquifer.

In addition to speed of cleanup, it is also necessary to keep a supply of oxygen and nutrients in the aquifer for the bacteria to use. Oxygen has a limited solubility, 7–10 mg/liter, depending on the temperature. Each pass of the water can supply only a limited amount of oxygen. Therefore, highly oxygenated water is recirculated through the aquifer as fast as possible. It is possible to add oxygen directly to the aquifer by placing intermediate wells between the central wells and the recharge wells, and bubbling air into the standing water of the well (6, 7). In this way, groundwater passing through the intermediate wells is supplied with oxygen.

It is not a good idea to place too much nutrients in the aquifer at one time. Both organic contaminant and nutrient should be used up at the end of an in-situ cleanup of an aquifer. Part of the operations of an in-situ cleanup must be the testing of the water drawn from the aquifer for residual nutrients. A minimal amount of nutrients should be applied on each pass when the aquifer is treated.

The second difference with the water flow in an in-situ treatment is that the water being recharged to the aquifer is not completely clean. The water entering the aquifer must have bacteria and nutrients. Figure 4–7 shows the zone of influence of the drawdown by the central well. The recharge water should not be placed outside of this zone of influence. The bacteria and, more likely, the nutrients could be considered a contaminant downflow in the aquifer.

The only way to insure that the water placed in a recharge well does not get out of the zone of influence is to place less water in the recharge well than was drawn from the central well. The zone of influence extends all around the central well. The plume travels in one direction from the well. The recharge well is placed at the end of the plume. While the recharge water will reduce the amount of water taken from

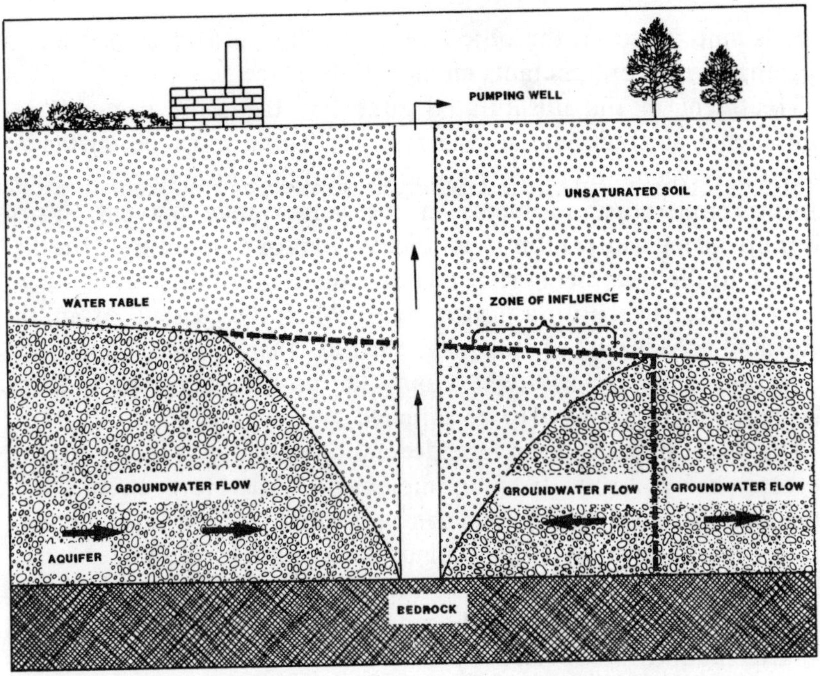


Figure 4-7. Zone of influence on an aquifer from a central well.

the opposite flow side of the central well, there will still be some water from the upflow side. The design engineer will have to work closely with the hydrogeologist to determine the correct ratio of the two flows. The excess water will have to be discharged off site or used as part of the cleanup of the unsaturated zone.

**The Unsaturated Zone.** The design engineer must either get bacteria, nutrients, and oxygen to the affected area in the unsaturated zone or flush the area of its contaminants. Most contaminations of aquifers are a result of material being released above the aquifer in the unsaturated zone. If the cleanup is limited to the aquifer, the contaminants still in the unsaturated zone can be a source of future contamination. In-situ cleanup techniques can also be applied to the unsaturated zone. The design criteria are the same as those for in-situ cleanup of an aquifer, with two exceptions. First, the upper layers of the unsaturated zone may not require any water pumping as part of the cleanup. Second, the objective in the unsaturated zone may be flushing of the contaminants along with in-place destruction.

Treatment in the unsaturated zone has the same bacterial requirements as treatment in the aquifer. There must be sufficient bacteria, oxygen, and nutrients, and the correct environmental conditions. Temperature problems in the winter may prevent in-situ cleanup during winter months in some regions of the country. One difference in the treatment criteria is that there must also be sufficient water present in the unsaturated zone. When the cleanup of an unsaturated zone is part of an aquifer cleanup, there is always sufficient water available. In cases where only the unsaturated zone is being cleaned, water may have to be supplied.

The upper three to seven feet, depending upon the type of soil, can be reached by normal farming implements. Therefore, these upper layers of the unsaturated zone can be treated by land-farming techniques. Land farming is the in-place destruction of organic material with bacteria. These techniques have been well established for the treatment of oily waste from the petroleum industry. Recently, these same techniques have been used on land contaminated with hazardous waste. A prime example is the cleanup of wood-treating sites by land farming. Bacteria have been successful in completely removing pentachlorophenol and a range of other phenol-based compounds. It is not the purpose of this book to discuss land farming in detail. However, land farming should be considered as part of an in-situ cleanup of an aquifer.

Unsaturated zones can also be activated by flowing water through the contaminated sections. Once again, water with bacteria, oxygen, and nutrients is applied to the unsaturated zone. When part of an aquifer cleanup, the applied water is collected at the aquifer. Application of water onto the unsaturated zone can be a replacement or supplement to the use of recharge wells. The collected water should be tested to insure that the proper environmental conditions exist in the unsaturated zone.

One limit to applying the water to the unsaturated zone is that the zone is made up of small soil particles. These particles can act as a filter for the bacteria. Even with large-particle soil, the bacteria will tend to move slower than the water and its soluble components. Below-surface injection of the water can help to alleviate this restriction. Complete distribution of large amounts of bacteria will still be difficult to accomplish.

A second property of the bacteria will help to reach all of the contaminants in the unsaturated zone. The contaminants in the unsaturated zone have adsorbed onto the soil particles. The bacteria will produce enzymes that will bring these compounds into solution. This action will help flush the contaminants from the unsaturated zone. This material can then be collected and used to grow more bacteria in aboveground tanks for further application to the unsaturated zone.

**Aboveground Equipment.** All in-situ cleanups will require aboveground equipment. Land farming needs farming equipment to spread the bacteria and nutrients and to provide mixing and oxygen. Aquifer cleanups will require a minimum of pumps and mix tanks for the supply of bacteria, nutrients, and oxygen. In addition, most aquifer and unsaturated-zone cleanups will require a treatment tank above ground. This treatment tank will be used for four main functions:

1. Produce bacteria
2. Reduce organic content of the water
3. Add oxygen to the water
4. Add nutrients to the water

Bacteria are the workhorse of an in-situ cleanup. Large quantities of bacteria are required for the process. Natural bacteria and commercial products are sufficient for the cleanup of the upper layer of the unsaturated zone. For lower layers of the unsaturated zone and the aquifer itself, however, these represent good seed sources only. A timely cleanup requires more bacteria than can be grown in place.

A standard biological treatment system is set up above ground. The system can be a lagoon, activated-sludge, or fixed-film system, depending on the amount of bacteria required. Bacteria will produce at a rate of between 0.10 and 0.25 lb of bacteria per pound of organic consumed. Any of the biological treatment systems discussed in the first section of this chapter are very effective in growing the required bacteria. The choice between the systems will be made mainly on the amount of bacteria required. Lagoons, which produce the highest quantity of bacteria per pound of organic, are good when the recharge water is to be applied to the soil surface.

Lagoons produce too high a concentration of bacteria when a well recharge is to be used, because the bacteria can clog the well. Activated-sludge or fixed-film systems with a clarifier will both discharge lower concentrations of bacteria. Settled bacteria in the clarifier can be added to the recharge water to obtain the desired bacterial concentration.

The food for the bacteria is readily available. The central well will be in the center of the plume, and the contaminants will be at their highest concentration at this point. The water drawn from the well is sent to the treatment tank, and the bacteria grow on the contaminants. This will produce bacteria that are actively degrading the compounds found in the aquifer. The same holds true for compounds flushed from the unsaturated zone.

The reaction tank accomplishes the second function simultaneously with the first function. The bacteria use the contaminants found in the groundwater for food. As they use the compounds, the organics are removed from the water. The recharge water has the bacteria in it, but not the original contaminants. The recharge water should not have a high concentration of the contaminants in it.

Although in-situ treatment will lower the cost of the project and speed its completion, the rate of biological destruction is always faster in an aboveground tank. Any contaminants that can be destroyed above ground should be. Also, the bacteria will be "hungry" for more food if they have consumed all of the food available. Removal of the contaminants will also minimize the oxygen requirements below ground.

One final reason for removal of the contaminants above ground is that the engineer will want to limit the amount of bacteria that are grown below ground. Once again, the aquifer and the unsaturated zone are made up of small particles. Too many bacteria grown below ground will tend to clog the subsurface structures. In-place reactions are best for low concentrations of contaminants, residual cleanup of high concentrations of organics, and flushing of organics from hard-to-reach places.

The final uses of the aboveground reaction tank are to supply nutrients and oxygen. The nutrients are added directly to the tank.

Enough nutrients must be added to satisfy the requirements of the biological reaction within the tank and to have a residual for the recharge water. The same is true for oxygen. The discharge concentration of oxygen from the reaction tank should be near the saturation level.

## REFERENCES


1. Bouwer, E. J., and McCarty, P. L. Modeling of Trace Organics Biotransformation in the Subsurface. *Ground Water,* **22**:4–433 (1984).
2. Rittmanm, B. E., and Brunner, C. W. The Nonsteady-State-Biofilm Process for Advanced Organics Removal. *Journal Water Pollution Control Federation,* **56**:7–874 (1984).
3. Kincannon, D. F., et al. Predicting Treatability of Multiple Organic Priority Pollutant Wastewaters from Single Pollutant Treatability Studies. 37th Purdue Industrial Waste Conference, May 1982.
4. Freeman, R. A., et al. Air Stripping of Acrylonitrile from Wastewater Systems. *Environmental Progress,* **3**:1–26 (1984).
5. Lee, M. D., and Ward, C. H. Biological Methods for the Restoration of Contaminated Aquifers. National Center for Groundwater Research, 1983.
6. Yaniga, Paul M. Groundwater Abatement Techniques for Removal of Refined Hydrocarbons. *Hazardous Wastes and Environmental Emergencies Proceedings* (March 1984).
7. Jhaveri, V. "Bio-Reclamation of Ground and Groundwater by the GDS Process" Groundwater Decontamination Systems, Inc.

# 5
# Treatment Methods for Inorganic Compounds

The main inorganic contaminants found in groundwater include the following:

- Heavy metals
  - Cadmium
  - Trivalent chrome
  - Hexavalent chrome
  - Copper
  - Lead
  - Mercury
  - Nickel
  - Silver
  - Zinc
- Arsenic
- Nitrates
- Total dissolved solids
- High and low pH

Inorganic contaminants in groundwater have not had as much attention as the organic contaminants in the past few years. There are several reasons for this. First, the ground, the unsaturated zone, has a limited ability to treat and remove these contaminants from a spill, so they never reach the aquifer. Most soils have an ion exchange capacity. A heavy metal moving through the soil will be exchanged with a cation in the soil and be removed from the spill. Anaerobic zones in the soil can biologically transform nitrates into nitrogen gas. The soil also has a natural ability to neutralize a high or low pH.

Inorganic compounds are also not used as often as organic compounds for industrial purposes. Industrial plants do not put heavy-metal solutions into large storage tanks. Pipelines are not used to transfer these compounds. Heavy metals in their pure state are not soluble in water. Even when stored as a salt, which is soluble in water, they are in a solid form. There is no equivalent for inorganic compounds to gasoline storage tanks, oil pipelines, solvent storage tanks, etc.

The largest source of heavy-metal contamination is leachate from abandoned sites. Legal landfills, illegal landfills, and old mines are probably the major source of heavy metals. Ten years ago, the regulations did not anticipate the problems caused by improper placement of heavy metal below ground.

The largest source of arsenic, nitrates, and total dissolved solids is agriculture. These sources represent low-concentration material being introduced into the ground. Agriculture is also only one source of contaminants for organic contaminants, mainly pesticides.

Finally, inorganic compounds are usually found in low concentrations in aquifers. There is no odor or color at these concentrations, and the public does not readily recognize the presence of low concentrations of inorganic compounds.

Low concentrations of nitrates, however, have been known for a long time to cause methemoglobinemia in infants. Recently the public has recognized that other low concentrations of inorganic compounds are also not acceptable. This, combined with the cleanup of abandoned sites, has made the inorganic area more important. The U.S. EPA has ongoing major projects in the area of inorganic contaminants in drinking water. The EPA will put these findings into new drinking water regulations in the near future.

This chapter will cover the following methods for removal of inorganics from groundwater:

1. Chemical addition
2. Removal of suspended solids
3. Ion exchange
4. Reverse osmosis, electrodialysis, distillation

A major advantage of developing treatment methods for inorganics in groundwater is that all of the methods described below can be tested in the laboratory. The results can be accurate enough to develop a preliminary design of the full-scale system. Biological treatment and stripping both require pilot plants to develop the necessary data, and carbon adsorption requires a column test for accuracy. The laboratory test for inorganic compounds can also be performed in a short period of time—days, versus the weeks and months that it takes to get good data for organic treatment methods. All of the methods listed below should be tested at several different concentrations and pH's. All water is slightly different, and although dosages and pH's are recommended in the text, the optimum conditions will have to be found for each groundwater.

## CHEMICAL ADDITION

### pH Adjustment

There are two main purposes for pH adjustment in the treatment of groundwater. The first is the adjustment of the groundwater to a neutral pH of around 7. Water that is to be discharged to a receiving stream must have a pH of between 6 and 9. The second purpose is the precipitation of heavy metals. Although the precipitation normally includes an adjustment of pH, other chemical addition is necessary when chemicals such as arsenic and hexavalent chrome are removed. This topic will be covered in the next section.

In wastewater treatment, the preferred method of pH adjustment is to use two waste products of opposite pH to adjust the pH of both. Groundwater will not have a secondary source for pH adjustment. Therefore, the main methods available for adjustment of an acid water are

1. Passing water through a limestone bed
2. Mixing water with lime slurries
3. Adding castic soda, NaOH
4. Adding soda ash, $Na_2CO_3$

The main methods for adjustment of an alkaline water are

1. Bubbling carbon dioxide in the water
2. Adding a strong acid, HC1, $H_2SO_4$, etc.

It is very rare for a groundwater to be too alkaline. The main reason for adding acids to a groundwater is for pH readjustment after the water has been raised to a high pH in order to precipitate a metal. A strong acid will be the normal method for this adjustment, because an industrial plant must be nearby in order for carbon dioxide to be cost effective.

The more common case of pH adjustment is the addition of alkaline materials. A limestone bed can be used when the purpose is to adjust the pH to the neutral range, around 7. Mountain streams that have been affected by abandoned mines often use large blocks of limestone to provide a constant pH adjustment. There are several volunteer groups that carry limestone in backpacks to remote sites in order to protect the trout fishing in these mountain streams. Limestone is only good for adjustments to neutral pH, and on relatively clean water.

The main reason for pH adjustment is to remove heavy metals from the water. The pH will normally have to be raised above 7 to remove the metals, especially when the metals are held in solution by a chelate. Figure 5–1 summarizes the solubilities of the various heavy metals as a hydroxide precipitate. Lime or caustic soda are necessary to reach these pH's.

The main differences between the two compounds are that lime is less expensive and more difficult to use. Lime costs about 40–60% less than caustic. Actual prices will depend heavily on transportation costs, so specific prices should be obtained for each project. Lime is normally fed as a hydraded lime slurry. The material is stored dry and slurried before it is mixed with the acid groundwater. Lines can become clogged easily, and great care should be taken to keep the slurry in motion.

Depending on the final pH, lime can take up to a 30-min reaction time to be completely used. The slurry should be introduced to the acid groundwater in a completely mixed tank. The residence time in the tank should therefore be 30 min. This slow reaction time will make

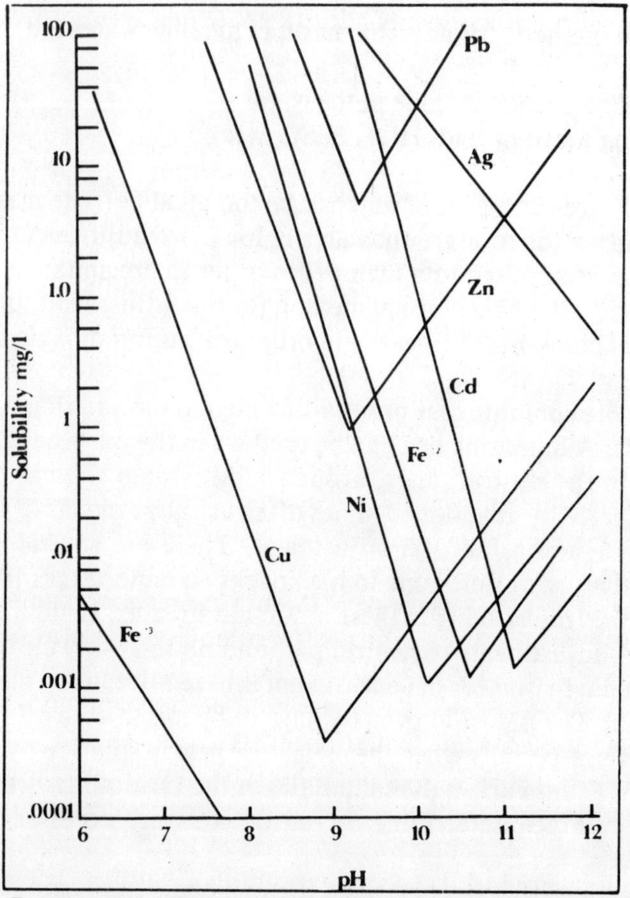


**Experimentally determined solubilities of metal hydroxides.**

Figure 5-1. Solubilities of metal hydroxides at various pH's. *(Courtesy of Graver Water.)*

pH control more difficult. Lime will also form more sludge than neutralization with caustic soda.

Caustic soda can be delivered and stored as a liquid. The reaction time is very fast with caustic soda. The reaction tank should still be completely mixed, and about 5-10 min residence time is sufficient. Caustic soda can be used at several different concentrations. This can be an advantage for small streams that are near neutral pH. Short-term—one to two years—groundwater treatment projects will probably use caustic soda for pH adjustment. Long-term projects may be able to take advantage of the low cost of lime.

## Chemical Addition Before Precipitation

Not all metals will precipitate upon an increase in pH. If the metals are being kept in solution by a chelate, they will not follow the solubility curves in Figure 5–1. In addition, iron in the ferrous state and chromium in the hexavalent state will not precipitate at high pH's. Arsenic is another inorganic compound that is not removed by a simple increase in pH. All of these compounds require a chemical addition before they can be precipitated.

One way to remove metals when they are being kept in solution is to precipitate them as a sulfide precipitate as opposed to a hydroxide precipitate. The solubility is still dependent on the pH, but in general, the metal sulfide is less soluble than a metal hydroxide. The effluent concentration of the metal will be less, and in the case of metals such as lead and zinc, the range of pH's that the reaction can be run is increased. Figure 5–2 summarizes the solubility of metal sulfide compounds.

Although the theory of metal sulfide precipitation looks good, the practical application of the technology is difficult. Short-term groundwater projects will probably not use this technology unless there is an industrial plant nearby that is already using the technology. Long-term projects may want to consider the technology if there is a reason that very low metal concentrations are required in the effluent from the treatment system.

Iron is not a toxic metal. However, it can cause problems with processes, pipes, equipment, and final use of the water. Iron is found in many groundwaters naturally. The technology for removing iron in the ferrous state and the ferric state are well established. Ferrous iron has to be oxidized to the ferric state. In the ferric state, iron is not soluble above a pH of 7. Oxidation occurs readily at pH 7.0–7.5. The water must simply be aerated at this pH and ferrous iron will convert to ferric iron. The methods for transferring oxygen discussed in the biological treatment section of Chapter 4 are all applicable to the oxidation of ferrous iron.

Hexavalent chrome is a toxic heavy metal. It is only used in in-

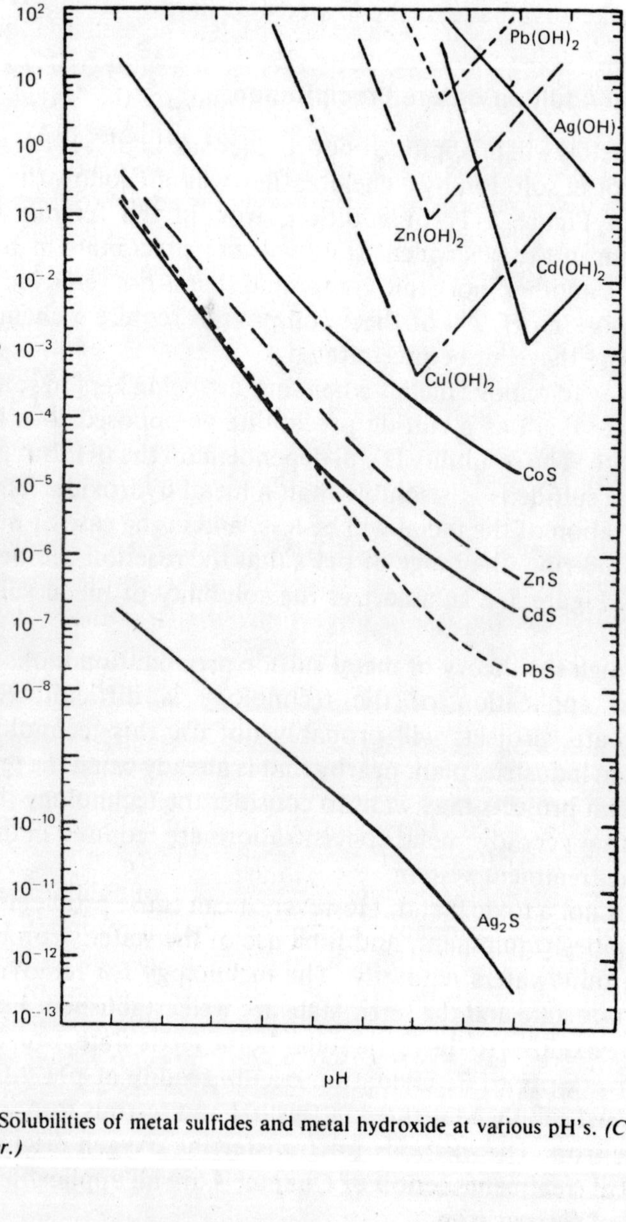


Figure 5-2. Solubilities of metal sulfides and metal hydroxide at various pH's. *(Courtesy of Graver Water.)*

dustrial plants and is not found naturally in groundwater. Like ferrous iron, it is soluble in water at high pH's. The methods for treating hexavalent chrome have been developed for industrial wastewater. The hexavalent chrome must be reduced to the trivalent state. Trivalent chrome is not soluble at high pH's (Figure 5-1). The standard reduction treatment technique is to lower the pH of the water to 3 or below. Next, a chemical reducing agent is added to the water. The most common reducing agent is sulfur dioxide, but sodium bisulfite, metabisultfite, hydrosulfite, or ferrous sulfate can be used. The pH is then raised and the trivalent chrome precipitates.

Hexavalent chrome reduces readily but not at a fast rate. The longer that the hexavalent chrome is in the ground and the more contact the spill has with other material and water, the more likely it will naturally be reduced to the trivalent state. However, when hexavalent chrome is found in the groundwater, at least a 20–30 min residence time should be used for the reduction reaction in the treatment system. As with all inorganic treatment methods, the reaction time for reduction of the hexavalent chrome should be established in laboratory tests. When hexavalent chrome is found in groundwater, it is most likely that a local industrial plant is responsible for the release. That plant will probably have the necessary treatment system for the reduction of the chrome.

One final inorganic chemical that requires chemical addition is arsenic. Arsenic requires the formation of a floc in order to be removed from the water. Increasing the pH with lime will remove some of the arsenic. The same increase in pH with caustic soda will not remove the arsenic, due to the lower amount of solids formed. The most efficient way to remove arsenic is to add iron, in the ferrous or ferric state, at a pH of between 5 and 6, and then to increase the pH to 8-9 with lime. Iron:arsenic ratios should be around 8:1. Ratios above or below that ratio will decrease the removal efficiency.

## REMOVAL OF SUSPENDED SOLIDS

### Flocculation

From the addition of various chemicals and the adjustment of pH, the groundwater now has all of its inorganic chemicals in a nonsoluble form. All of the material is now suspended solids that must be removed from the water. These suspended solids are all heavier than water. However, these solids will not be removed simply by placing the groundwater in a quiescent tank and allowing the particles to settle.

For example, assume that all of the particles have a 2.65 specific gravity. The particles that have a diameter of 0.1 mm, about the size of fine sand, will take 38 sec to settle 1 ft. Particles with a 0.01-mm diameter, about the size of silt, will take 33 min to settle the same 1 ft. Finally, particles with a diameter of 0.001 mm, about the size of bacteria, will take 55 hr to settle 1 ft. Colloidal particles would take years to settle the same distance (1).

The sizes of the particles generated by the chemical addition and pH adjustment will range from colloidal to fine sand. The specific gravity will be lower than 2.65 and will vary depending on the metal precipitated and the chemicals used. As can be seen, a simple tank will not remove all of the suspended solids. The small particles must be brought together before they will settle from the water. This process is called flocculation.

There are two basic steps to flocculation. First, the particles all have the same charge on their surfaces, usually a negative charge. This charge is what keeps the particles separate. The first step is to neutralize this charge so that the particles can come into contact. Once the particles are in contact, they will not separate unless subjected to high shear forces.

Inorganic coagulants—lime, alum, ferric chloride, and others—can be used for this purpose. When the pH is raised with lime, both pH adjustment and coagulation occur. The other compounds will be used mostly when the final use of the treated groundwater is drinking. However, lime will probably be the preferred method for most groundwater treatment systems.

Once the charge is neutralized, the second step is to grow the particles to even larger sizes. This step is called flocculation. The particles require gentle mixing so that they come into contact with other particles, but not with so much force that the particles are broken apart. Flocculated solids should never be run through a centrifugal pump, because the sheer forces in the pump would easily tear apart the fragile floc.

The best way to increase the size of the particles depends upon the concentration of the suspended solids. The higher the concentration, the more contacts that will occur between the particles. At low concentrations, the particles are so far apart that the gentle mixing would have to continue for a very long period of time in order for all of the particles to contact the other particles.

One method to improve the efficiency of flocculation is to introduce an organic flocculating agent. These flocculants are made up of high-molecular-weight polymers. The various ends of the polymers attach to different particles, bringing them together. One to 5 mg/liter of these compounds is usually sufficient to increase the settling rate by twofold or threefold. Another advantage of the polymers is that the solids capture is improved.

Although the polymers will improve flocculation, the main problem to overcome is the effect of concentration on the frequency of particle contact. The equipment design must solve this problem. The type of clarifier used in the groundwater treatment system will depend upon the concentration of solids.

### Settling Equipment

There are several types of settler designs that can be employed to remove suspended solids. This book will review four designs:

1. Clarifier/thickener
2. Flocculating clarifier
3. Solids contact clarifier
4. Lamella

These four designs are reviewed with the purpose of identifying the major components of each design. On short-term groundwater treatment projects, the design engineer may not want to spend the money on a preconstructed settling tank. An existing tank or a portable tank can serve as a quiescient tank in which solids can settle. The engineer should use the description of the four different types of settlers to be sure that the tank will have the components necessary to do the job.

A clarifier/thickener is the simplest of the four designs. The three other designs can be considered clarifiers with certain components added. A clarifier/thickener must perform several tasks. First, the water with the suspended solids must enter the tank without causing turbulence in the tank and reagitating already settled solids. Next, the water must be evenly distributed throughout the tank in order to make maximum use of the surface area of the clarifier. The supernatant must then be collected and removed from the tank. Finally, the solids must be thickened and removed from the tank.

Figure 5–3 depicts a standard "center feed" clarifier. The water enters into the center feed well. The feed well protects the contents of the clarifier from the energy contained in the incoming water. If the flow is to be introduced into the side of the tank, a plate or half-sphere (outside facing out) should be placed at the inlet pipe. This will serve the same energy dissipation function as the feed well.

The key to water distribution in the clarifier is the influent section and the effluent collection section. The influent must be introduced as specified above. The effluent must be collected over as large a surface area as possible in the tank. This is accomplished through the use of a saw-toothed weir on top of a trough. The overflow rate should be limited to 13 gpm/ft of weir length. Lower levels are always better. The weir should be level throughout the tank in order to insure proper distribution of the water in the tank. A flat weir can be used, but it is more difficult to adjust to the correct height throughout the clarifier. The troughs are connected to a single pipe which exits from the clarifier.

The solids settle to the bottom of the tank. With flocculating solids, it is important to have sufficient depth for the floc to grow inside the clarifier. The depth also allows the solids to concentrate before they are removed from the tank. The less liquid that goes with the solids, the less material that has to be sent for disposal. The solids must be moved to one location in order to be removed from the tank. The clarifier depicted in Figure 5–3 uses scrapers on the bottom of the tank to push the solids toward the center of the tank. The solids are collected in a sludge well and exit the tank through a pipe. An alternative for scrapers is to slope the bottom of the tank at 60° to a central point. This is called a hopper bottom clarifier, and is viable only for small designs.

Clarifiers should be designed at 0.2–0.4 gpm/ft$^2$ of surface area. The maximum diameter of a portable clarifier is 12 ft (wide load on a

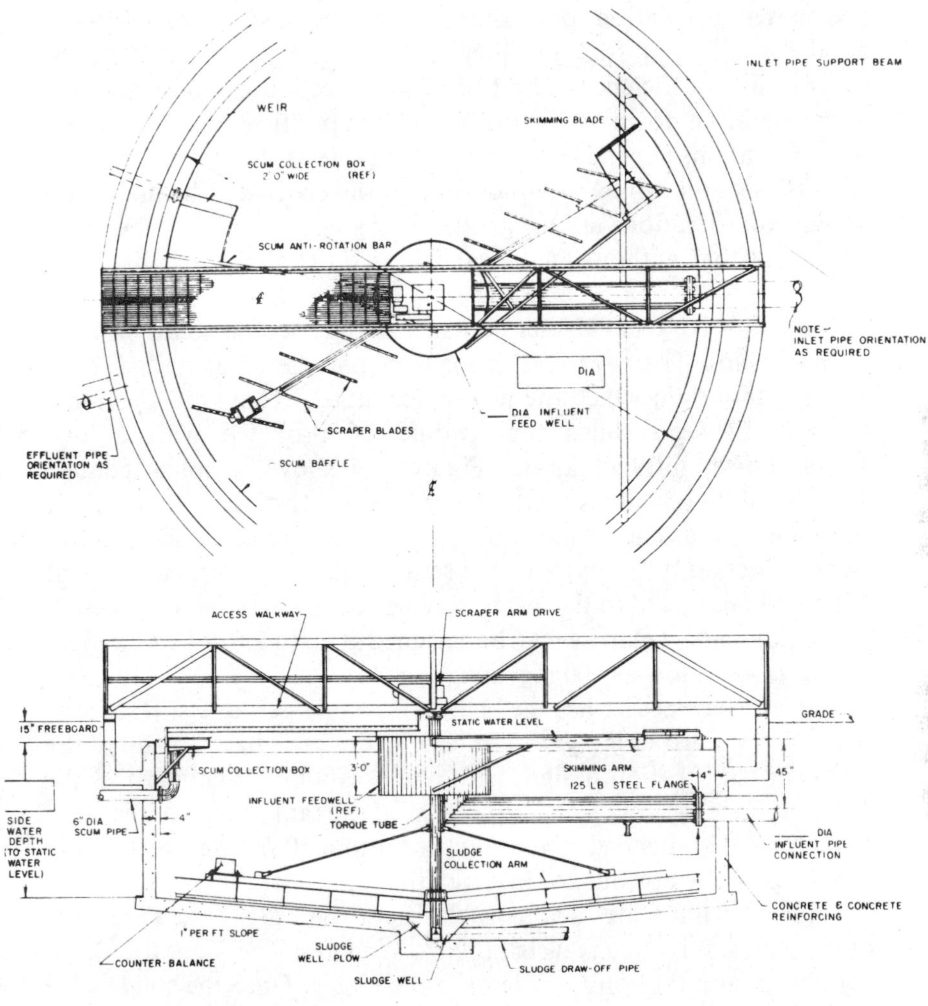


Figure 5–3. Center feed clarifier. *(Courtesy of Smith & Loveless, Inc.)*

truck). Therefore, the maximum flow rate of a portable clarifier is about 40 gpm. Multiple units or a site-erected clarifier will have to be employed for flows above this level.

A flocculating clarifier has all of the same components as a regular clarifier. The only difference is that the influent well is expanded in size. The influent well has gentle mixing also added to it, and it becomes a flocculation zone inside the clarifier. A standard clarifier is good for solids of 1000 mg/liter and above. Below that level, the clarifier loses efficiency. The flocculation section extends that efficient operation down to 500 mg/liter. The specific concentration for both operations depends upon the settling rate of the solids. The flocculation section can be set up outside of the clarifier. The important design considerations are for gentle mixing and transfer of the water to the clarifier without any shear forces on the floc that has been formed.

The solids contact clarifier contacts already settled solids with incoming solids. The key to settling is large particles. The settled solids act as a core onto which the new solids attach. There are also more contacts between solids than would be provided with a low-concentration influent alone. Figure 5–4 shows a solids contact clarifier. Two advantages are realized with this design. First, lower suspended solids can be introduced to the clarifier and removed efficiently. Second, the recycle of the solids makes more efficient use of the chemicals added to the water. Savings of 20–30% can be realized. The solids contact clarifier can be designed at up to 1.0 gpm/ft$^2$ and influent solids as low as 100 mg/liter.

The lamella clarifier has the highest flow rate per tank surface area of all of the clarifier designs. The lamella design is based on all of the basics of clarification design, with one exception. It emphasizes the fact that once the solids hit the bottom of the tank, they are removed from the water. Instead of putting the bottom 10 ft from the top, the lamella puts the bottom 2–4 in. from the top.

Figure 5–5 illustrates a lamella clarifier. The water enters the sides of the plates. It is equally distributed to all plates. The water travels up the plates, and the solids settle onto the plates. Once the solids have settled on the plates, they are removed from the water. The solids on the plate continue down the plate as in a hopper bottom clarifier. The slope of the plates is 45°–60°.

What makes the lamella design so powerful is that the theoretical settling area is the projected area of the plates, Figure 5–6. The projected area of all of the plates is additive. The resulting projected area can be 10 times the liquid surface area of the tank. Therefore, up to 10 times the liquid flow can be applied to the same size tank.

The solids that are removed with the plates fall off the plates and

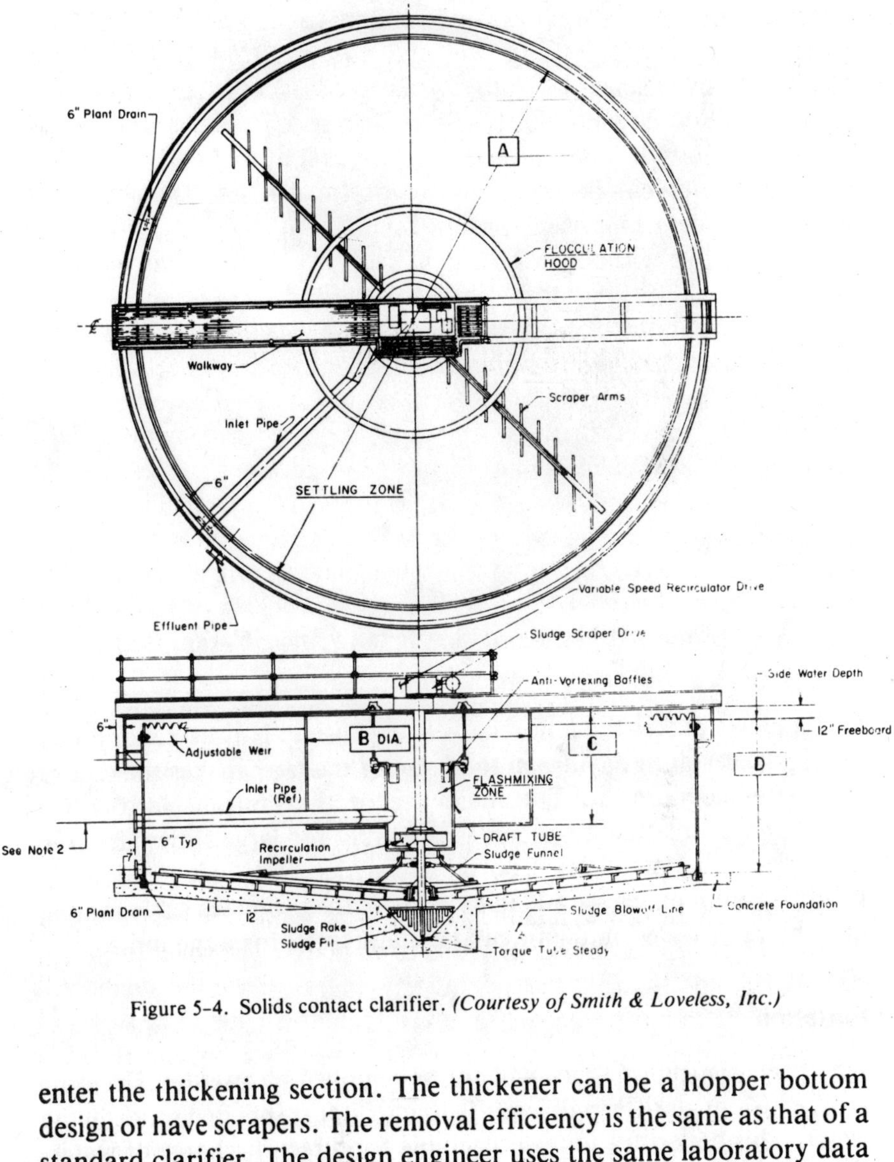


Figure 5-4. Solids contact clarifier. *(Courtesy of Smith & Loveless, Inc.)*

enter the thickening section. The thickener can be a hopper bottom design or have scrapers. The removal efficiency is the same as that of a standard clarifier. The design engineer uses the same laboratory data to design the lamella but uses the projected area of the plates for the surface loading rate.

For flows that are small enough for a portable clarifier, 40 gpm, the lamella is probably not economical. When the tank is small, the cost

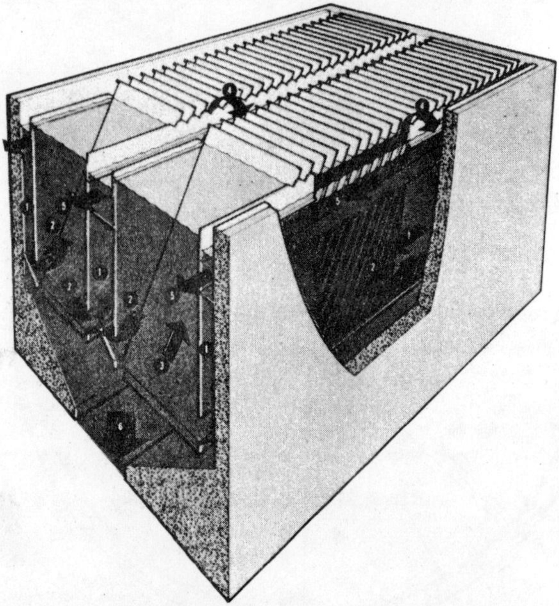

Figure 5-5. Lamella clarifier. *(Courtesy of Graver Water.)*

of the plates makes the unit relatively expensive. However, for larger flows, the resulting savings in total size of the tank makes the cost of the plates economical. The main use of the lamella clarifier for groundwater treatment systems is probably for large flows. Even in large sizes, the lamella is a portable clarifier. The unit increases in length as more plates are added to increase the projected surface area, and the 12-ft wide-load limit can be maintained for large units.

## Filtration

All of the suspended solids will not be removed by a settler. Even the most efficient clarifier will leave 5–10 mg/liter suspended solids in the water. This concentration is acceptable for direct discharge of water. However, when the water is to be used for drinking water or process water, even this low concentration is too high. The suspended solids must be removed by other technology. This is also the case when the original concentration of suspended solids is very low. If the initial

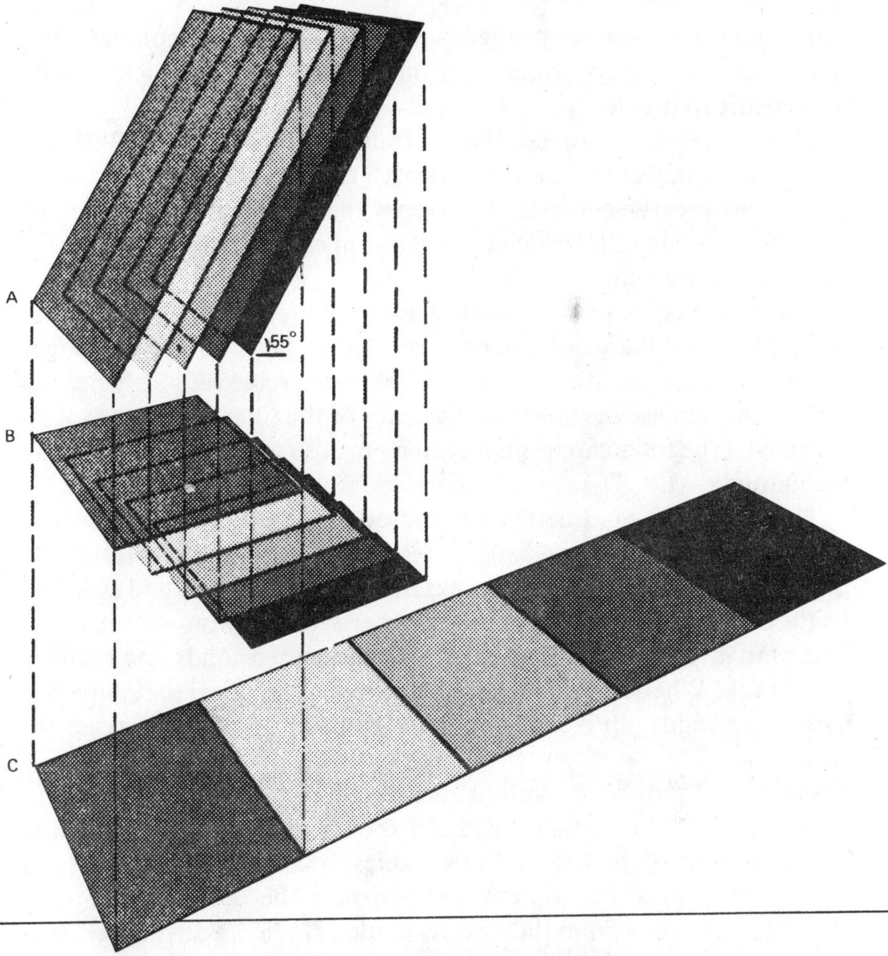


Figure 5-6. Theoretical settling area of a lamella clarifier. *(Courtesy of Graver Water.)*

concentration of heavy metal is low, then the resulting suspended solids from chemical addition and pH adjustment will also be low.

Low suspended solids must be filtered from the water. The simplest form of filtration is to pass the groundwater through a bed of sand. The suspended solids attach to the sand particles, and the water continues through the bed. It is very important for the design engineer to realize that a sand filter does not strain the solids from the water. A sand filter removes suspended solids by solid-solid contact. The suspended solids in the water come into contact with a particle of sand and attach to it.

This is the same process that is found in flocculation. With the tremendous amount of sand in the path of the water, the chances of contact are greatly increased. So, even with very low concentrations of suspended solids, the particles will come in contact with a sand particle and be removed.

The other basics of flocculation are also important for sand-filter operation. The charge on the particle must be neutralized. The larger the particle, the greater the chance of its being removed. Polymer addition will increase the removal efficiency of the filter. The removal efficiency of the filter can be improved from 50% to 90% by use of these techniques.

The filter is also cleaned based on flocculation theory. Clean water flows up through the sand bed at a fast rate. The sand bed is fluidized, and the turbulence in the bed breaks off the particles that had attached to the sand. The water takes the resuspended solids out of the filter. The sand is heavier and larger than the suspended solids and remains in the tank. When an increase in shear forces is required to remove the suspended solids, air is added to the sand bed in order to increase the turbulence.

Figure 5–7 shows a downflow, gravity, dual-media filter. In a sand filter, there is 2 ft of sand instead of 1 ft of sand and 1 ft of anthracite. The water enters the top and hits a splash plate, which prevents the water falling onto the sand from disturbing the bed of sand. Sand filters can also be run in the upflow mode. There are advantages and disadvantages to this flow pattern, but most sand filters today are the downflow variety. The distance from the top of the filter chamber to the top of the sand should provide enough room for a 50% expansion of the sand bed during backwashing.

Next comes the sand bed. When the bed is backwashed, the smallest particles end up on top of the bed. As solids are filtered out of the water, they build up in the sand. The smaller the sand particles, the faster the solids build up. The solids will finally fill most of the void spaces in the sand bed. When this happens, the filtration method switches from solid/solid contact to straining. Straining of solids from the water takes a great deal of force, and the water builds up on

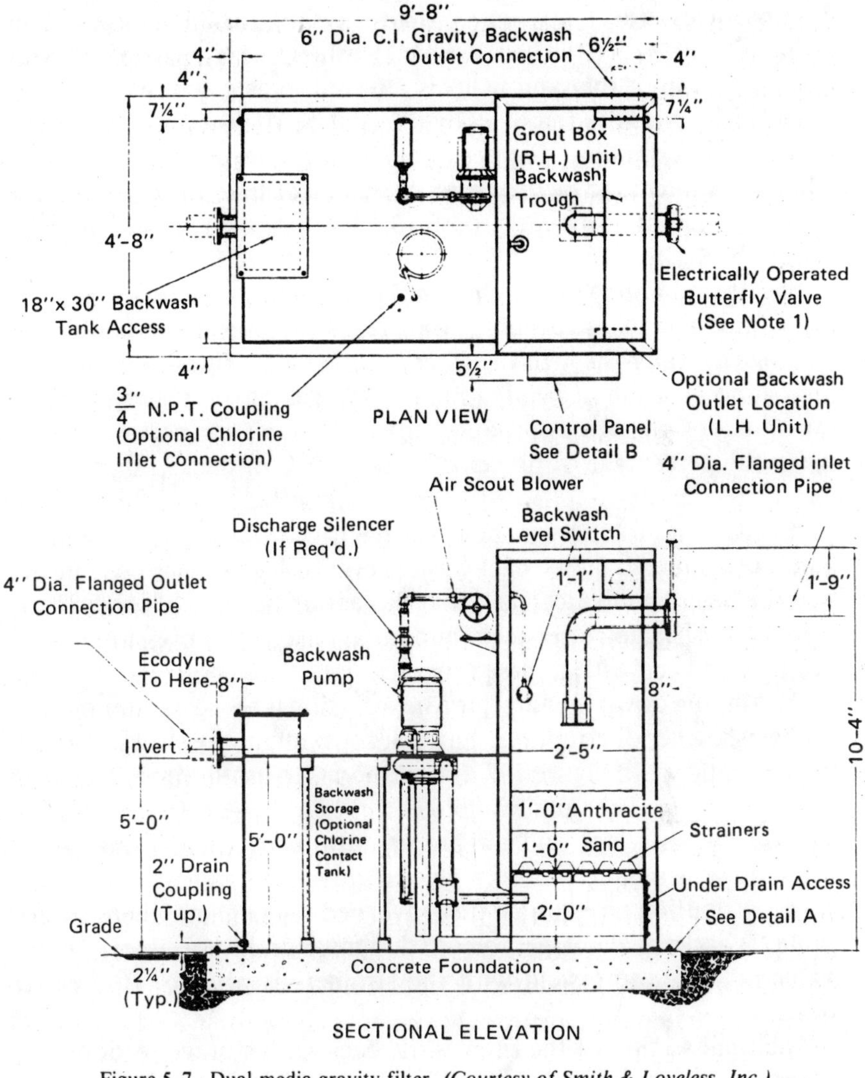


Figure 5-7. Dual-media gravity filter. *(Courtesy of Smith & Loveless, Inc.)*

top of the bed. Gravity filters are normally set up with a water column above the bed to force the water through the sand. These columns can be up to 15 ft above the sand.

There are two main ways to extend the use of the sand bed. The first of these is to place a bed of coal above the sand bed (Figure 5-7). The coal particles are slightly larger than the sand, and they are lighter than the sand. When the beds are backwashed, the larger coal particles end up on top of the sand. This is referred to as a dual-media filter. Suspended solids are removed by the coal before they ever get to the sand, thus extending the length of time between backwashing. Anytime a filter is to be used after a biological system, or with a high concentration of suspended solids (50–100 mg/liter), the dual media should be used.

The second way to extend the time between backwashes is to seal the filter tank and to use a pump to increase the pressure available for forcing the water through the filter bed. Pressure filters are good for high-concentration suspended solids, 100–250 mg/liter. The filter run must be long enough to produce sufficient water for backwash. The lower the percentage of processed water used for backwash, the better. A maximum of 10% of the processed water should be used for backwash, and the design should strive toward 2–5%. On the other hand, the filter should be backwashed every 24–48 hr. This will insure that the bed remains clean and that there is no buildup of solids in the filter bed. The choice between sand, dual media, and pressure should be made to keep within these ranges.

Continuing down the filter, the next section is the sand support and backwash water distribution. There are two main ways to accomplish these functions. In Figure 5–7, the method is to use strainers on top of a support plate. The sand cannot get through the slots in the strainer. The backwash water and air are equally distributed to all of the strainers. The second way is to place a gravel bed under the sand and have distribution pipes inside the gravel bed. Once again, there are advantages and disadvantages to both designs. For the portable, groundwater market, the system with the strainers is probably the better system.

The final section of the filter is the backwash storage section. The water to be used for backwashing should be relatively clean of suspended solids. The water that has already been processed by the filter should be used for backwash. Backwash water should flow at 15 gpm/ft$^2$ of filter surface area for 5 min. This water can be stored beside the tank and pumped at the necessary rate, or the water can be stored above the filter and flow back through the filter by gravity. Air, if desired, should flow at 3–8 scfm/ft$^2$ of surface area for 5 min. A blower should be used for supplying the air.

One final advantage of the sand and dual-media filters is that their operation can be set up to be automatic. A pressure or time setting can be used to initiate the backwash cycle. No operator attention is necessary for the proper operation of these filters.

## MISCELLANEOUS METHODS

### Ion Exchange

The ion exchange properties of soil have been recognized since the 1850s. Since that time, there have been many improvements in the materials that can exchange an ion in the water for an ion on the solid-phase exchange material. The largest use of ion exchange technology today is the use of synthetic resin beads for softening of home potable water.

Ion exchange is basically the exchange of an ion with a high ion exchange selectivity for an ion with a lower selectivity. Any divalent ion will usually have a higher ion exchange selectivity than a monovalent ion. Table 5-1 summarizes the selectivity for different ions on a variety of ion exchange resins. Calcium, which is divalent, will replace sodium, which is monovalent, at an exchange site on an ion exchange bead. This is the basis of water softening. The calcium ion, hardness, exchanges with the sodium ion on the ion exchange resin. The calcium is removed from the water and the water has lost the ions that make it "hard."

The resin is regenerated by passing a high concentration of sodium ion through the ion exchange bed. All reactions go in both directions. The calcium will exchange with the sodium, but at the same time the sodium will exchange with the calcium. The difference is that the rate

**TABLE 5-1. Ion Exchange Resin Selectivity.**[a]

| RESIN | SELECTIVITY[b] |
|---|---|
| Strong acid | $Li^+$, $H^+$, $Na^+$, $NH_4^+$, $K^+$, $Rb^+$, $Cs^+$, $Mg^{2+}$, $Zn^{2+}$, $Cu^{2+}$, $Ca^{2+}$, $Pb^{2+}$ |
| Weak acid | $Na^+$, $K^+$, $Mg^{2+}$, $Ca^{2+}$, $Cu^{2+}$, $H^+$ |
| Strong base | $F^-$, $OH^-$, $H_2PO_4^-$, $HCO_3^-$, $Cl^-$, $NO_2^-$, $HSO_3^-$, $CN^-$, $Br^-$, $NO_3^-$, $HSO_4^-$, $I^-$ |
| Weak base | $F^-$, $Cl^-$, $Br^-$, $I^-$, $PO_4^{3-}$, $NO_3^-$, $CrO_4^{2-}$, $SO_4^{2-}$, $OH^-$ |

[a] From Paterson, J. W. "Wastewater Treatment Technology." Ann Arbor, Mich.: Ann Arbor Science, 1978.
[b] Increasing selectivity left to right.

of reaction for the calcium to replace the sodium is much faster than the opposite reaction. However, a high concentration of sodium ions in the water, relative to the calcium ions, will drive the reaction in the opposite direction. The ion exchange resins are put back into their sodium form and are ready for removing further hardness from the water.

All of the heavy metals present in water are in the divalent or trivalent state, with the exception of hexavalent chrome. A simple home, sodium ion exchange unit will remove all of these compounds. However, the process is expensive, and the regeneration brine, with the heavy metals, will still have to be disposed of off site. These two problems severely limit the use of ion exchange for large quantities of heavy metals. The best use of ion exchange is for very low concentrations and for final treatment before potable use.

Heavy metals are cations, positively charged. Ion exchange can also be used to remove anions, negatively charged. Chlorides, nitrates, sulfates, etc., can be removed by anion exchange resins. Hexavalent chrome is, in fact, removed by anion resins.

All ions can be removed by ion exchange. Sodium ions can be removed by ion exchange resin in the hydrogen form. Hydrogen ions exchange with the sodium ions in the water. Sodium has a higher exchange potential than hydrogen. Combining anion ion exchange resin in the hydroxide form with cation exchange resins in the hydrogen form will remove all of the ions in the water. The remaining hydrogen and hydroxide ions combine to form water.

This process is used to make ultrapure water for high-pressure boilers. The same process could be used to treat groundwater when the contaminant is general dissolved solids. Once again, this is a very expensive process. Normally, an aquifer would be abandoned instead of cleaned of dissolved solids. When treatment is necessary, reverse osmosis or distillation would be the preferred method.

## Reverse Osmosis

Reverse osmosis (RO) uses semipermeable membranes and high pressures to force pure water through the membrane. The membrane rejects inorganic material and allows the passage of water. The separation is not perfect. Depending upon how the membrane is prepared, the salt rejection is anywhere from 50% to 99%. Staged systems can accomplish any required removal efficiency.

Low-molecular-weight organic compounds pass through the membrane at rates far above inorganics. Reverse osmosis is not a good technology for the removal of organic material.

Reverse-osmosis systems are readily available. They are expensive to run, due to the high pressures required, 100–250 psig. The pH must be maintained between 5.5 and 7.5 to protect the membrane. Great care must also be taken to insure that no precipitation occurs in the RO module.

### Electrodialysis

Electrodialysis is a combination of membrane technology and ion exchange technology. Electrodialysis uses ion exchangers in membrane form. The driving force across the membrane is provided by electric current. The ions are thus removed from the water and pass through the membrane, attracted by the opposite electric charge on the other side of the membrane.

The advantages of the system are that the residence time controls the amount of dissolved solids removed, and that the system can be run continuously with no regeneration required. The disadvantage of the system is that the water must carry an electric current. The cleaner the water, the more resistance to the current, which increases the cost of operation.

### Distillation

Distillation is the evaporation of the water molecule followed by recondensation. The inorganics do not evaporate with the water and are left behind. The condensate is purified water. The process requires heating of the water to increase evaporation rates and cooling of the airstream to condense the water vapor. Volatile organics will evaporate and condense with the water.

The cost of heating and cooling the water can be very high. However, new technology in the area has produced multiple-effect distillation. Basically, this process uses the same energy several times in the process. Multiple-effect distillation has had broad applications on water desalination projects in the Middle East. Although the technology is readily available, the cost is still relatively high for groundwater treatment.

## REFERENCE

1. Powell, S. T. "Water Conditioning for Industry." New York: McGraw-Hill, 1954.

# 6
# Field Application of Design Methods

## INVESTIGATION OF SUBSURFACE GASOLINE CONTAMINATION *

Doug Gore
*Douglas Engineering*
*Concord, California*

### INTRODUCTION

This section presents the findings for the investigation of subsurface gasoline contamination at a service station in California.

Inventory monitoring records indicate that approximately 6000 gal of premium unleaded gasoline was lost from the valve at the sump pump of the premium unleaded underground storage tank (see plot plan, Figure 6–1). The investigation has been directed toward this spill. However, the investigation and monitoring well system are applicable to any other gasoline spill that may have occurred in the vicinity of the underground gasoline storage tanks.

The scope of services encompassed obtaining information pertinent to the situation; performing a surface grid hydrocarbon survey, planning and installing the monitoring well system; sampling and analyzing soil, water, and free hydrocarbons obtained from the wells;

---

*This section provided courtesy of Douglas Engineering.

measuring free hydrocarbon and water levels; and measuring the direction and magnitude of the groundwater flow.

## UTILITY AND TANK INFORMATION

The information that was searched includes state, county, municipal, and private documents regarding soil conditions, groundwater level and flow direction, safety and permit requirements, and utility locations and as-built drawings of the site and its underground tanks.

Neither the City Public Works Department, the Consolidated Fire District, nor the previous owner of the site have as-built drawings for the site and its underground tanks. Information from the owner and site observations have been incorporated into the plot plan, Figure 6–1 which shows three underground gasoline storage tanks. Two 8000-gal tanks, each 22 ft long and 8 ft in diameter, lie parallel to the street. The tank closest to the street stores super unleaded gasohol while the other, adjacent, 8000-gal tank stores the premium unleaded. A more recently installed 10,000-gal tank containing regular gasoline lies to the west of and perpendicular to the two original 8000-gal tanks.

The required distance by code between the tanks and the sides of the original excavations for the tanks was 1 ft. An asphaltic concrete patch over the location of the 10,000-gal tank is 10 ft across, indicating that the excavation for this tank had vertical sides. Backfill material for tank excavations is typically sand or gravel. The tops of the tanks are located approximately 3 ft below the ground surface. The sump pumps (square covers) and fill boxes (round covers) are typically located 10 in. in from the ends of the tanks, although this is not the case for the sump pump for the 10,000-gal tank which is located near the middle of the tank (see Figure 6–1).

Tract map information, which was obtained from the City Public Works Department, shows the locations of sanitary sewers in the vicinity of the site. A sanitary sewer line runs parallel and adjacent to the back property line. Observations inside a sanitary sewer manhole at the back of the site and a discussion with City Sewer Maintenance staff at the site confirm that the sewer line extends upslope at least as far as a nearby car wash. The flow line of the sanitary sewer is at elevation 64 ft, which is 4 ft below the manhole cover according to the preliminary design drawings for the sanitary sewer also obtained from

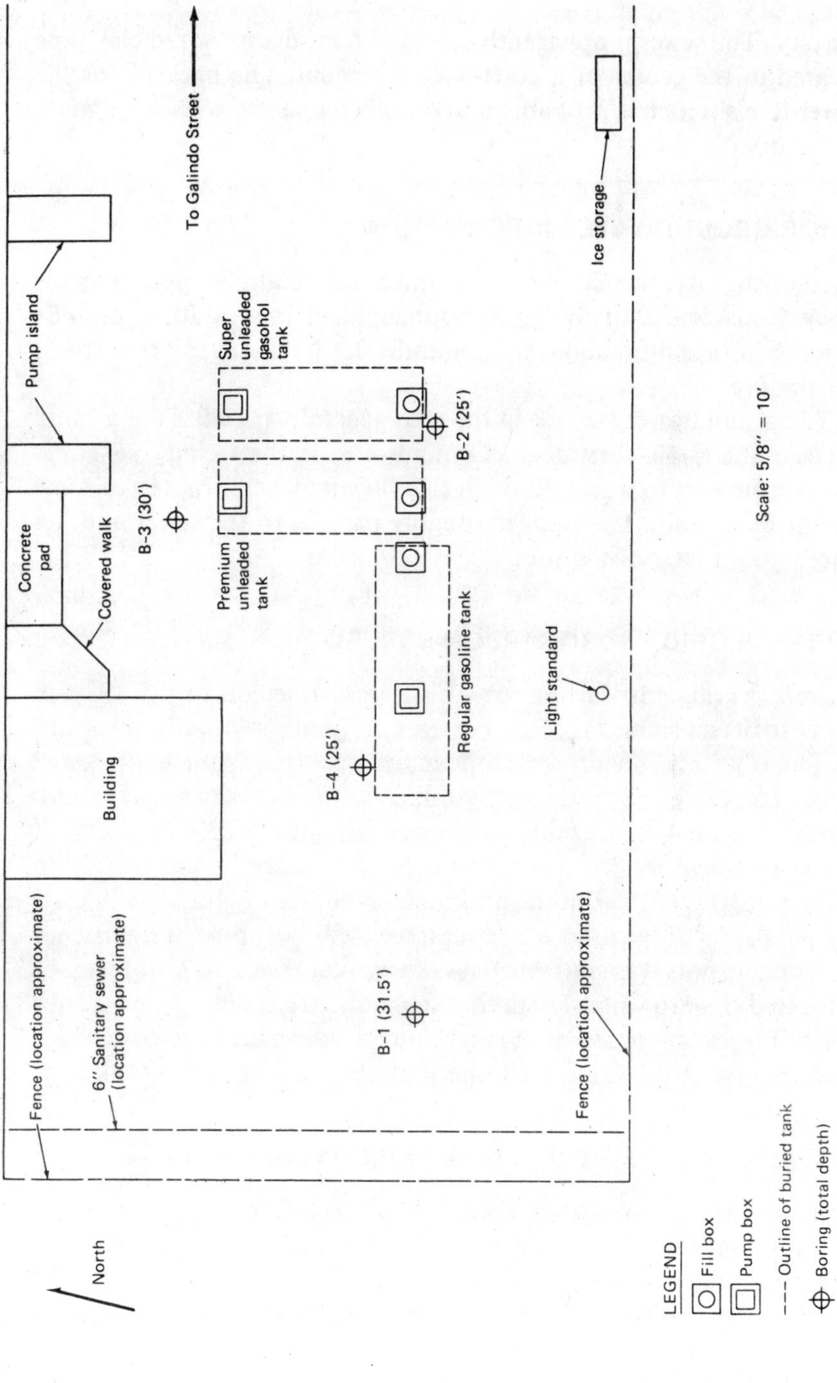


Figure 6-1. Plot plan for gasoline station.

the city. The sewer is apparently made of 6 in.-diameter red clay pipe located in the center of a 10-ft-wide easement. The backfill for the sewer line's trench is probably native soil compacted to 90% of maximum dry density.

## PRIOR GROUNDWATER INFORMATION

Neither the city, the County Department of Health Services, nor the State Department of Water Resources have been able to provide specific information about the groundwater level and soil conditions for the site.

The groundwater surface in the area generally parallels the ground surface and is 10–20 ft deep. Groundwater in the vicinity generally flows from east to west, although this flow may be disrupted to some extent by a fault. The fault is roughly parallel to the street and lies about 400 ft west of the site.

## SURFACE GRID HYDROCARBON SURVEY

A surface grid hydrocarbon survey was performed on July 6, 1984, in order to detect locations of concentration ("hot spots") of subsurface gasoline. With an air-driven hammer and a 1½-in.-diameter bit, small holes were driven 18 in. into the ground at 5-ft centers in a grid pattern around the area of the tanks as shown on Figure 6–2. A GasTechtor brand hydrocarbon "sniffer" was used to measure the concentration of hydrocarbon fumes in the air in each hole (see Figure 6–2). After a second round of readings was completed on July 9, 1984, the existence of two hot spots was confirmed. As shown on Figure 6–2, hot spot #1 is located close to where the premium unleaded gasoline leaked from a valve at the sump pump of the premium unleaded tank. Hot spot #2 is located next to the fill box for the super unleaded gasohol tank.

## INSTALLATION OF MONITORING WELL SYSTEM

After consideration of all information presented above, a monitoring well installation plan was devised so as to obtain information about soil and groundwater conditions, locate the free product (liquid gasoline layer) that can be recovered, monitor the extent of con-

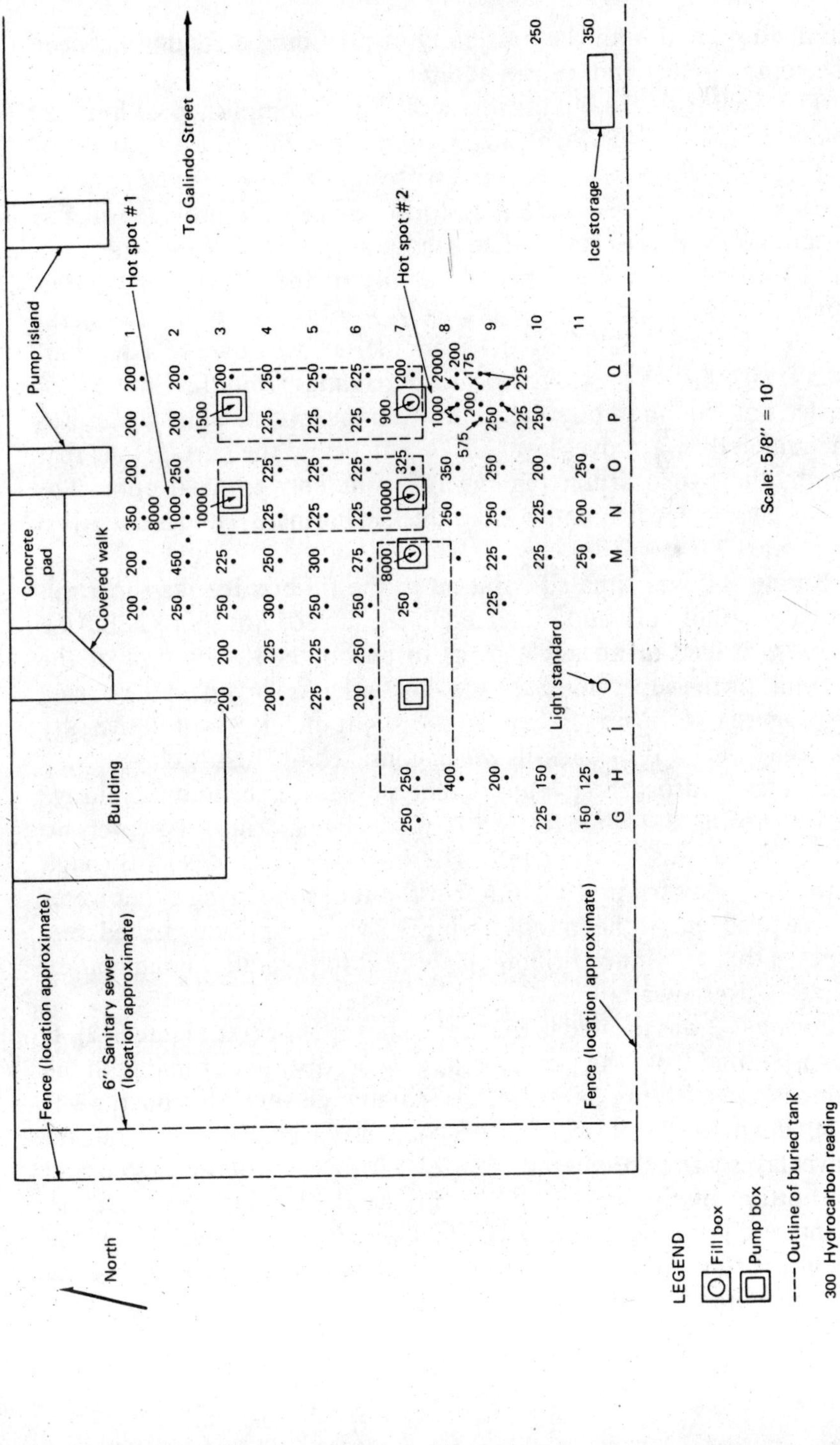


Figure 6–2. Hydrocarbon sampling for gasoline station.

tamination, and limit the possibility of providing a conduit between the contaminants and a clean aquifer.

The drilling of the monitoring wells was accomplished with truck-mounted continuous-flight auger equipment with the exception of boring B-4, which was completed with 8-in. hollow-stem equipment. Boring B-1 was located as a monitoring well and to provide soil and groundwater information. The well was set back from the tank backfill, in light of the high probability of the backfill being contaminated (see plot plan, Figure 6–2). Various clays and clayey silt were encountered in boring B-1 from the surface down to a depth of 24.5 ft (see Figure 6–3). Hydrocarbon readings from the GasTechtor device were in the 90 ppm range down to the level at which water was encountered in a gravel layer at 24.5 ft below the surface. At that depth, the hydrocarbon reading rose suddenly to 4800 ppm. The gravel layer is approximately 2.5 ft thick in boring B-1 and is underlain by at least 3.5 ft of silty clay.

Boring B-2 was situated adjacent to the fill box for the super unleaded gasohol tank and is located in the area of hot spot #2. Boring B-2 was drilled at an angle so as to encounter the bottom of the backfill for the super unleaded gasohol tank. As shown on the boring log, Figure 6–4, boring B-2 encountered 4 ft of fill, 5 ft of native silty clay, and then 2 ft of backfill which consisted of slightly clayey, very fine sand. Hydrocarbon vapor readings exceeded the lower explosive limit in boring B-2 while the top 11 ft was being drilled; however, no free product was encountered. The boring was advanced through native silty clay from 11 to 16.5 ft without encountering either water or free product. At that point, boring B-2 was temporarily halted over concern that continued drilling might lead to possible contamination of clean groundwater.

Boring B-3 was located in the area of hot spot #2 (see Figure 6–2). It was separated from the tanks by the new product lines running to the pump islands. Boring B-3 encountered native clayey soils from the 3-ft depth down to 22.5 ft (see Figure 6–5). A water-bearing silty sand and gravel layer was encountered from 22.5 ft to 26 ft. Between 26 ft and the bottom of the hole at 30 ft, silty sand and silty clay were encountered. Although no free product was found in this boring, the gasoline vapor concentration in the air down the hole was above the explosive limit.

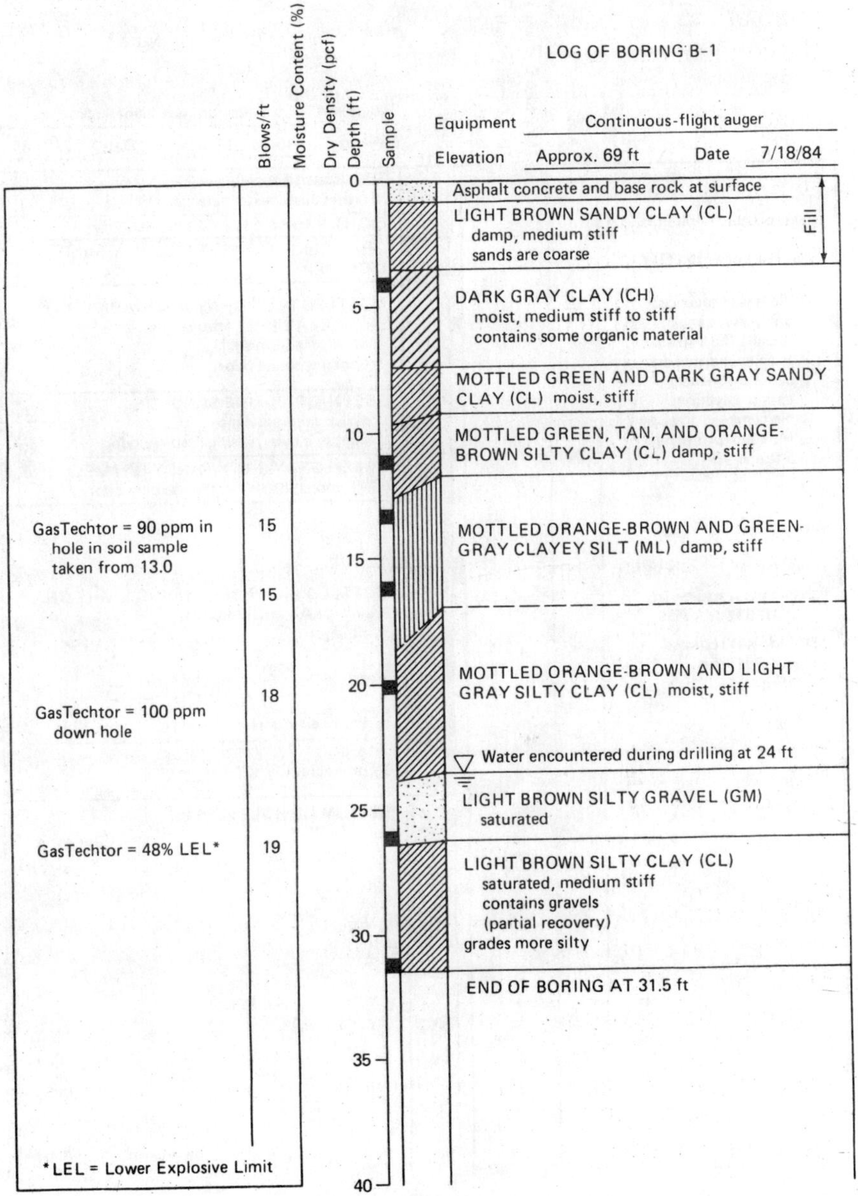


**Figure 6–3. Log of boring B-1.**

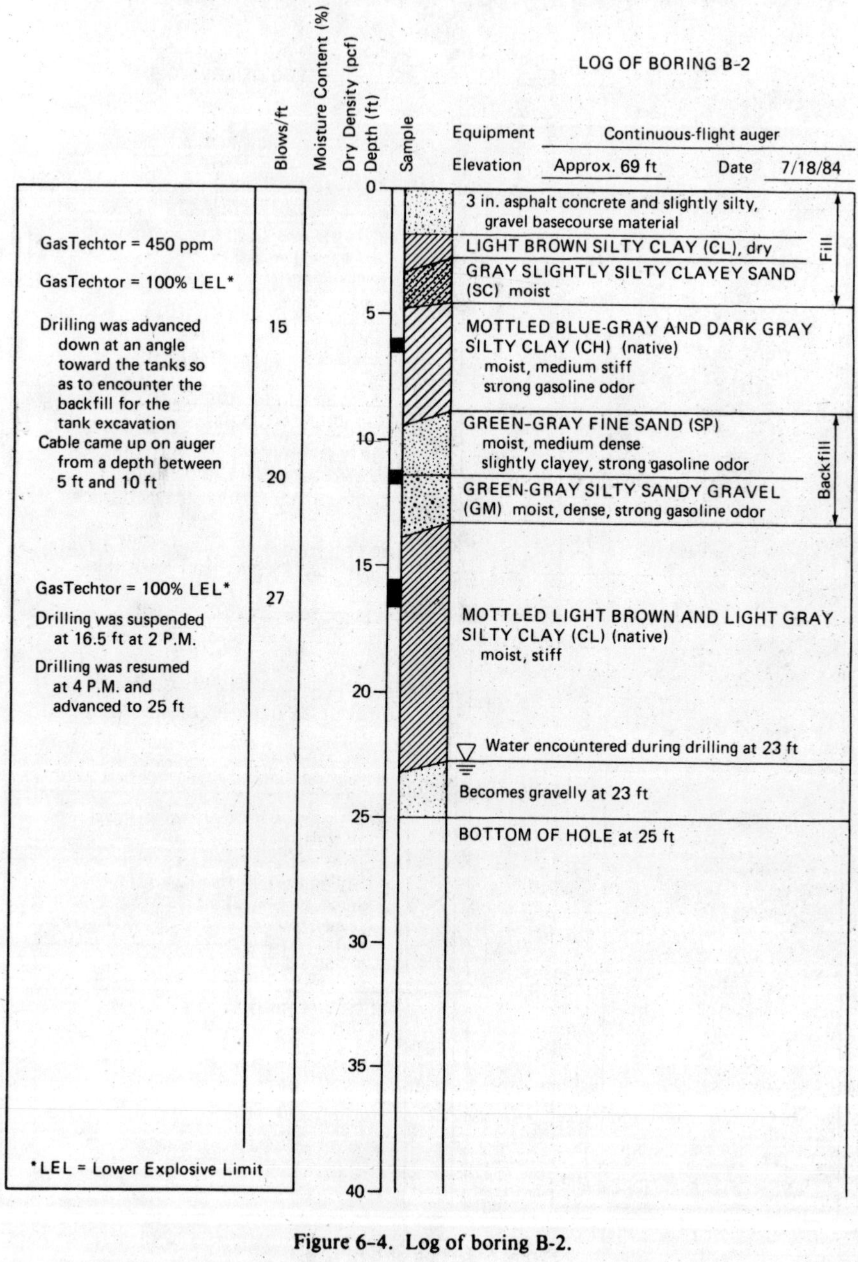


**Figure 6–4. Log of boring B-2.**

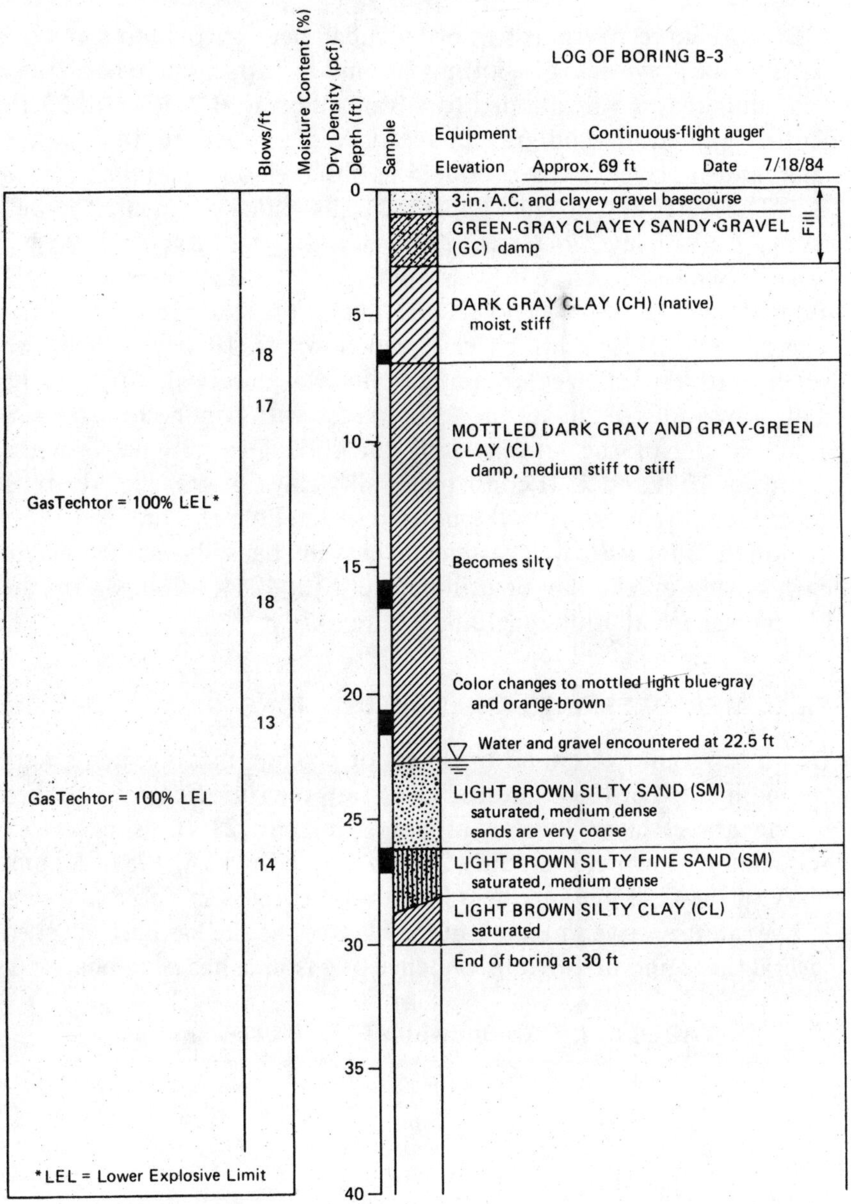


Figure 6-5. Log of boring B-3.

Because no recoverable free product had been encountered and the water in the gravel layer in borings B-1 and B-3 appeared to be already contaminated, it was decided to advance boring B-2 down to 25 ft. Slotted well casing and gravel backfill was installed, and two hours later, a ½-in. layer of free product was observed on top of the water in B-2. The water and product appeared to have flowed up into the well from the gravel layer located at 23 ft below the ground surface. Boring B-4, shown on Figure 6–6, was advanced into the gravelly backfill along the side of the 10,000-gal tank which stores the regular gasoline. The gravelly backfill which extended down to 10 ft was wet with water, but very little evidence of gasoline was observed. At 15 ft, the 6-in. continuous-flight augers were replaced with 8-in. hollow stems to avoid caving of the gravelly tank backfill. The soils between the depths of 10 ft and 23 ft consisted of silty clays. A gravelly layer was encountered from 23 ft to the bottom of the hole at 25 ft.

Slotted, 2-in. PVC well casing and gravelly backfill was installed in each boring on the day of drilling, July 18, 1984, as shown on the piezometer installation diagram, Figure 6–7.

## FIELD MEASUREMENTS AND OBSERVATIONS

Depth to groundwater and thickness of product measurements were taken on July 26, 1984. Groundwater flow measurements were taken in the gravel layer located at approximately 23 ft using a K-V Associates' flowmeter on July 25, 1984, and July 26, 1984. A summary of the field measurement results is presented in Table 6–1.

The sanitary sewer manhole at the back of the site has been checked several times and no obvious evidence of gasoline has been observed.

**TABLE 6–1. Groundwater Flow Measurements**

| BORING | DEPTH TO WATER (FT) | THICKNESS OF PRODUCT LAYER (IN.) | MEASURED IN GRAVEL LAYER: FLOW DIRECTION | MEASURED IN GRAVEL LAYER: FLOW VELOCITY (FT/DAY) |
|---|---|---|---|---|
| B-1 | 17.5 | 0.0 | Not measured | |
| B-2 | 18.5 | 0.5 | Not measured | |
| B-3 | 19.0 | 0.0 | N 85° E | 0.9 |
| B-4 | 18.0 | 3.0 | S 85° W | 0.33 |

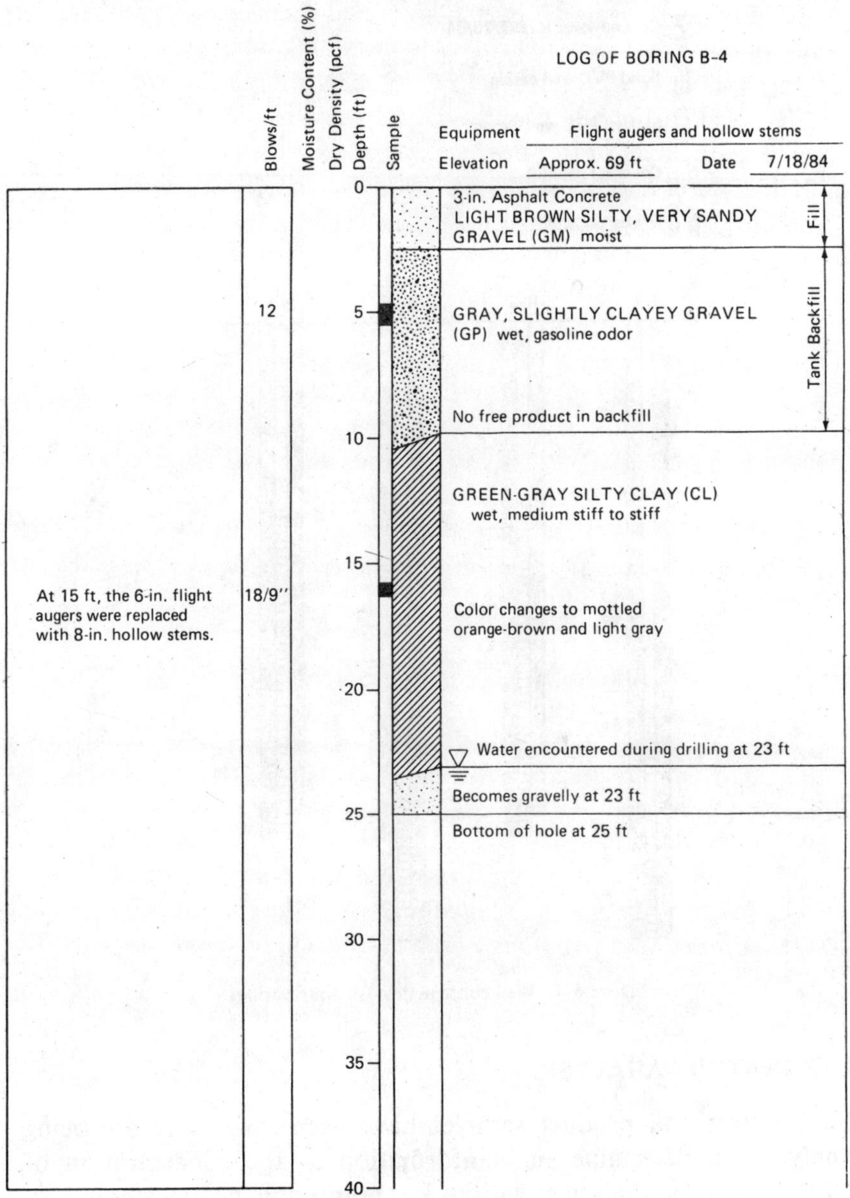


Figure 6–6. Lof of boring B-4.

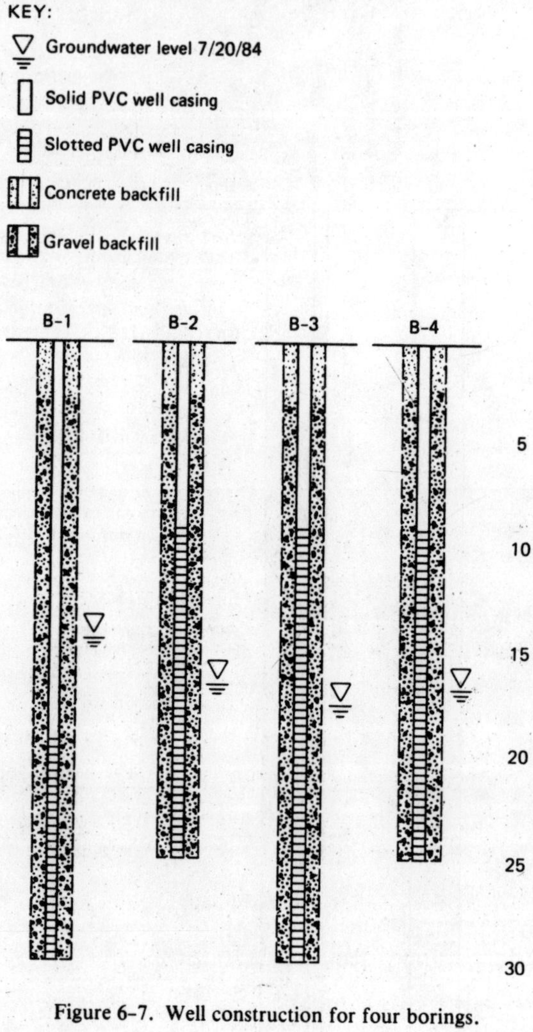


Figure 6–7. Well construction for four borings.

## LABORATORY ANALYSIS

Soil, water, and product samples have been taken and are being analyzed to determine such information as the concentration of hydrocarbons in the soil at various key points and in the water in borings B-1 and B-3, the type of gasoline in borings B-2 and B-4, and the permeability to gasoline of the soil below the tanks.

**TABLE 6-2. Laboratory Analyses**

| BORING | SAMPLE | TYPE OF ANALYSIS |
|---|---|---|
| B-1 | 20′0″ (soil) | Total hydrocarbon analysis |
| B-1 | 26′0″ (soil) | Hydrocarbon fingerprinting analysis |
| B-1 | 31′0″ (soil) | Total hydrocarbon analysis |
| B-1 | Water | Total hydrocarbon analysis |
| B-2 | 11′0″ (soil) | Total hydrocarbon analysis |
| B-2 | 16′0″ (soil) | Total hydrocarbon analysis |
| B-2 | Product layer | Hydrocarbon fingerprinting analysis |
| B-3 | 8′6″ (soil) | Total hydrocarbon analysis |
| B-3 | 16′0″ (soil) | Permeability to gasoline |
| B-3 | Water | Total hydrocarbon analysis |
| B-4 | 5′0″ (soil) | Total hydrocarbon analysis |
| B-4 | 15′4″ (soil) | Total hydrocarbon analysis |
| B-4 | Product layer | Hydrocarbon fingerprinting analysis and qualitative lead analysis |

Table 6–2 is a list of laboratory analyses being performed on selected samples.

The total hydrocarbon analyses provide gasoline concentrations in soil and water samples. The hydrocarbon fingerprinting analyses should be able to determine the types and concentrations of gasoline for the product layers above the groundwater in borings B-2 and B-4 and in the soil samples from the gravel layer in borings B-1 and B-3. The permeability to gasoline analysis will give an indication of the potential for the premium unleaded leak to have migrated through the soils underlying the backfill for the tanks. The qualitative lead analysis will determine if lead exists in the subsurface contamination.

## PRODUCT RECOVERY

Product reclamation was started October 5, 1984. A Douglas Engineering skimmer and depression unit were installed in boring 4 (B-4 at a depth of 25 ft. The unit was set to maintain a depression level of 24.5 ft and the free product recovery unit operated at a 24- to 23-ft depth. The discharge water tested at 330 ppm dissolved hydrocarbon concentration which was treated by a System 340-7 biodegradation system from Detox, Inc. The treated effluent was discharged into bor-

**TABLE 6-3. Velocity and Directional Readings**

| BORING | MEASURED IN GRAVEL LAYER FLOW DIRECTION | FLOW VELOCITY |
|---|---|---|
| B-1 | E 15° N | 0.5 ft/day |
| B-2 | Not measured | |
| B-3 | S 5° E | 1.5 ft/day |
| B-4 | Not measured | |

ing 2 (B-2) at a rate of 5 gpm. Product recovery from boring 4 (B-4) yielded about ¼ gal/min with pumping intervals at every 10 min.

New velocity and directional readings taken by the K-V Associates' GeoFlo Meter were as recorded in Table 6-3.

It is assumed that the system is stabilized and that removal of water from B-4 is creating a sufficient depression to affect water at B-1 and B-3, while creating a "flushing" effect by the discharge entering B-2.

### Future Activities

The system will remain in place until recovery conditions change or winter rains change the nature of the aquifer.

## AIR STRIPPING FOR A MUNICIPAL GROUNDWATER SUPPLY*

Kevin Sullivan
*Hydro Group, Inc., Linden, New Jersey*

Air stripping has proven to be a cost-effective method of treating groundwater contaminated with volatile compounds. Its efficiency and economy have been proven at many hazardous waste sites and contaminated public water supplies, such as the case of Acton, Massachusetts.

*This section provided courtesy of Hydro Group, Inc.

The Water District of Acton (located outside Boston) traditionally had little treatment of their groundwater supply. However, in December 1978, the district lost 40% of its water supply when two wells were shut down after several organic chemicals—trichloroethylene, dichloroethylene, methylene chloride, and benzene—were detected. The two wells (Assabet No. 1 and No. 2) each have a yield of 0.5 mgd, and are located in the 375-acre Sinking Pond aquifer.

The district set out to discover the source of the contamination. A year-long hydrogeological study, including over 100 test wells, determined the probable cause of contamination to be the waste disposal practices of a nearby chemical plant. The main sources were a landfill and waste lagoons located 3000 ft away from the wells. The study also found an extensive plume (10,000 ppb total hydrocarbons) to be within 1000 ft of Assabet No. 2 (Figure 6–8).

After initiating conservation measures and securing an additional water supply from a neighboring town, the water district then began investigating possible treatment alternatives for the Assabet wells. Concurrently, the legal aspects of the contamination came to court in April 1980, when the U.S. EPA filed the first suit in Massachusetts under the Federal Resource Conservation and Recovery Act (RCRA). Later, the water district filed a separate $3 million suit against the chemical company. The federal suit was settled by a consent degree in which the company agreed to assist in cleaning up the aquifer.

The water district chose activated carbon as its first method of treatment to meet their self-imposed maximum level of 1 ppb of any single organic, and no more than 5 ppb overall. In June 1982, water from Assabet No. 1 was pumped through two GAC adsorbers, each containing 20,000 lb of GAC. The carbon systems reduced the organic levels from an average of 42 ppb to below the district's standard of 5 ppb.

However, the high cost of carbon recharges quickly became a problem. Approximately every five months, a complete recharge of all 40,000 lb of carbon was necessary, which translated into an operating cost of $37,000. The carbon alone created an operating cost of $0.30/1000 gal. Additionally, the influent concentrations continued to rise, which would further shorten the life of the GAC.

At this point, Acton decided to investigate packed tower air stripping as a pretreatment. Hydro Group's Environmental Products Divi-

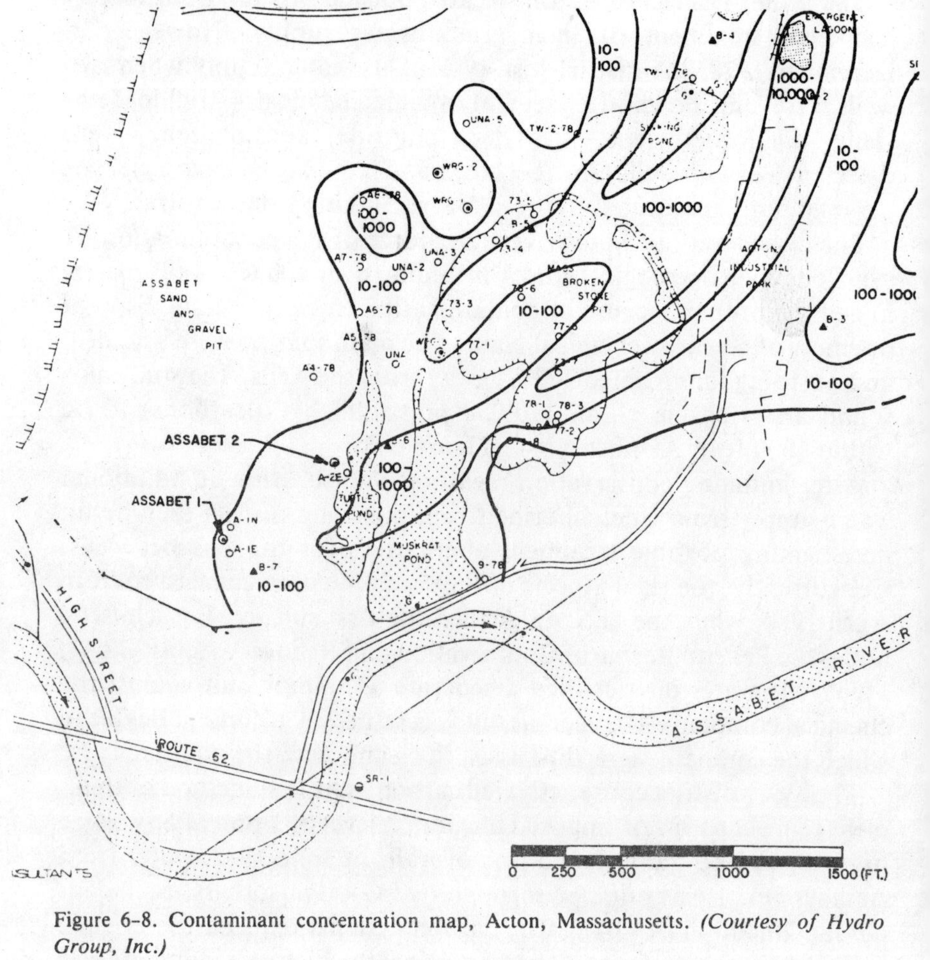


Figure 6–8. Contaminant concentration map, Acton, Massachusetts. *(Courtesy of Hydro Group, Inc.)*

sion was called in to run a pilot test and install a full-scale air stripper.

In June 1983, an on-site pilot test was conducted, using a small, mobile packed tower (Figure 6–9). Seven different runs were performed in a single day, examining the effect of liquid loading rates, air:water ratios, and packing height on the removal process. The rates and ratios for the test were based on previous tests conducted at similar sites; this allowed the scope of the test to be much more narrowly defined. Water loading rates varied between 10 and 30 gpm/ft$^2$, while air:water ratios ranged between 20:1 and 100:1. Samples were

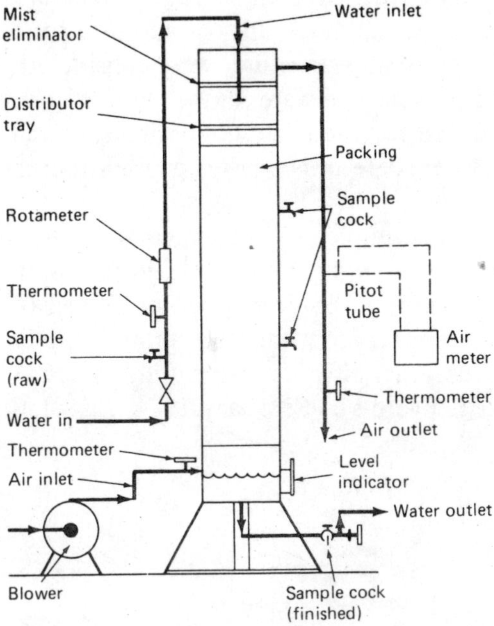


Figure 6–9. Mobile packed tower. *(Courtesy of Hydro Group, Inc.)*

collected after 10 and 15 ft of packing and sent to a laboratory, along with raw water samples for analysis.

Once the samples were analyzed, percentage removal and mass transfer coefficients were determined for each of the runs. With these mass transfer coefficients, and the aid of some iterative computer programs, a variety of tower configurations could be modeled. The column was designed to handle 700 gpm and to provide 95% removal of the VOCs. Since the removal rate required was so "low" (towers with removal rates as high as 99.9% are in operation), the hydraulic loading on the column was increased to nearly 30 gpm/ft$^2$, allowing for a reduction in the diameter of the column and an associated cost savings. The air:water ratio was kept to a minimum in order to conserve on the blower energy requirements.

The final design of the tower was a 66-in.-diameter × 28 ft-overall-high unit, designed to hold a packed bed of 2-in. Tripacks 20 ft deep. The 5-hp blower will supply 4700 cfm at 3 in. water gauge static pressure to achieve a 50:1 air:water ratio.

Once the process sizing was completed, the structural design began. Wind, snow, and earthquake loads, in addition to the "dead" and "live" loads found in any structure, were considered. The weight of the packing and the retained water in the column under the operating conditions were also factored into the design (Figure 6-10). The column thicknesses were determined based on the structural properties of 6061-T6 aluminum.

The column was completely fabricated in Hydro Group's New Jersey manufacturing facility. Meanwhile, in Acton, the water district installed a sump to mount the air stripper on, as well as booster pumps to take the treated water from the stripper and deliver it through the carbon adsorption units into the distribution system.

The unit was delivered in February 1984. Installation of the unit

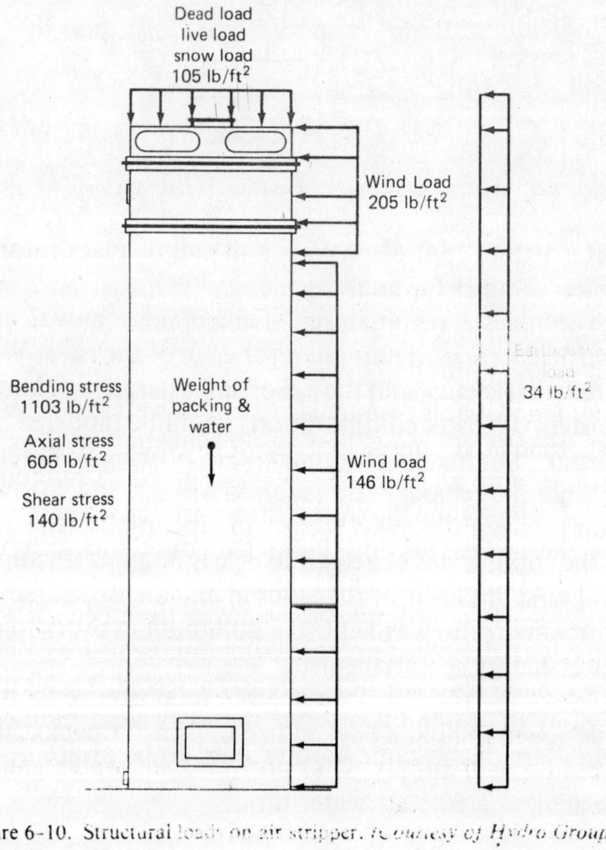


Figure 6-10. Structural loads on air stripper. (Courtesy of Hydro Group, Inc.)

took less than a day, using a small crane. When the unit was placed on line in April, the results were excellent. All organics were removed to less than 1 ppb. Removal percentages of 96–99% were recorded. These results are better than the predicted 95% due to the safety factors included in the full-scale design.

The capital cost of the air stripper was $31,000. Installation of the sump, a building to hold all of the electrical equipment, the pumps, and other miscellaneous equipment cost $109,000. Using 10% interest and a 20-year life, the yearly cost of capital equipment was $16,444. Electricity cost $0.06/kwh. The flow through the air stripper was 700 gpm. Total cost of air stripping the drinking water for Acton, Massachusetts, was $0.053/1000 gal.

Carbon life will be greatly extended by the air stripper at a fraction of the cost of carbon. Water district officials are very pleased with the installation, and plans are currently being developed to install another stripper at a recently contaminated well in the district.

## LOW-CONCENTRATION ORGANIC REMOVAL FOR A DRINKING WATER*

Mark H. Stenzel
*Calgon Carbon Corporation*
*Pittsburgh, Pennsylvania*

Groundwater is a valuable resource for potable water. A recent survey indicated that 36% of municipal potable water supplies rely on groundwater, serving an estimated population of 75 million in 1980 (1). Many of these municipalities draw groundwater from large aquifers, extending well beyond their legal boundaries and providing water for other communities and other uses.

In many cases where groundwater sources have been found to be contaminated with dissolved organic compounds, nearby wells drawing from the same aquifer are not currently affected. This is due to the slow movement of groundwater and the associated plume of contamination. Users of the same aquifer need to be aware of the con-

*This section provided courtesy of Calgon Carbon Corporation.

tamination, however, and should conduct periodic testing to determine whether the contaminants have reached their wells.

A typical case of aquifer contamination, and gradual spread of the problem, occurred in northern New Jersey in late 1980. New Jersey is a typical groundwater use area, as 47% of its public water supplies use this resource (1). In this case, Rockaway Township, a small community in Morris County, discovered that all three of its public wells were contaminated with trichloroethylene (TCE). In November 1980, the township installed a granular activated-carbon system to remove the contaminant.

Rockaway Borough, a nearby community of 7800 (Figure 6–11), became concerned about the quality of the drinking water it was drawing from the same aquifer and initiated a volatile organic scan program. Initially, only traces of tetrachloroethylene (PCE) were detected in one well, which was then shut down to prevent contamination of the entire supply. This testing confirmed earlier findings by the New Jersey Department of Environmental Protection, which was conducting tests in all towns near Rockaway Township using the same aquifer.

Within three months, the PCE level at this well had increased to 554 ppb, and the borough established a water use restriction program to cope with the loss of one of its wells. Borough officials realized that the contaminant plume could be moving into their well field and would eventually reach other operating wells.

"At this point . . . the Council and I were deeply concerned," recounted Borough Mayor Robert Johnson. "Closing down the one well only brought us temporary relief; we decided we had to find a way to remove the contaminant from our water" (2).

The mayor and Borough Administrator Walter Krich, Jr., immediately began to explore options for a permanent solution to treat the contaminated groundwater. Prior to installation of the treatment system (described later), however, contamination had spread to the other two wells, with detection of TCE as well as PCE.

On February 28, 1981, the mayor advised the community to discontinue use of tap water for drinking or cooking and announced that the borough's entire water system was to be shut down. Until a treatment system was in place, an emergency drinking water supply was made available using water trucks supplied by the National Guard, as shown

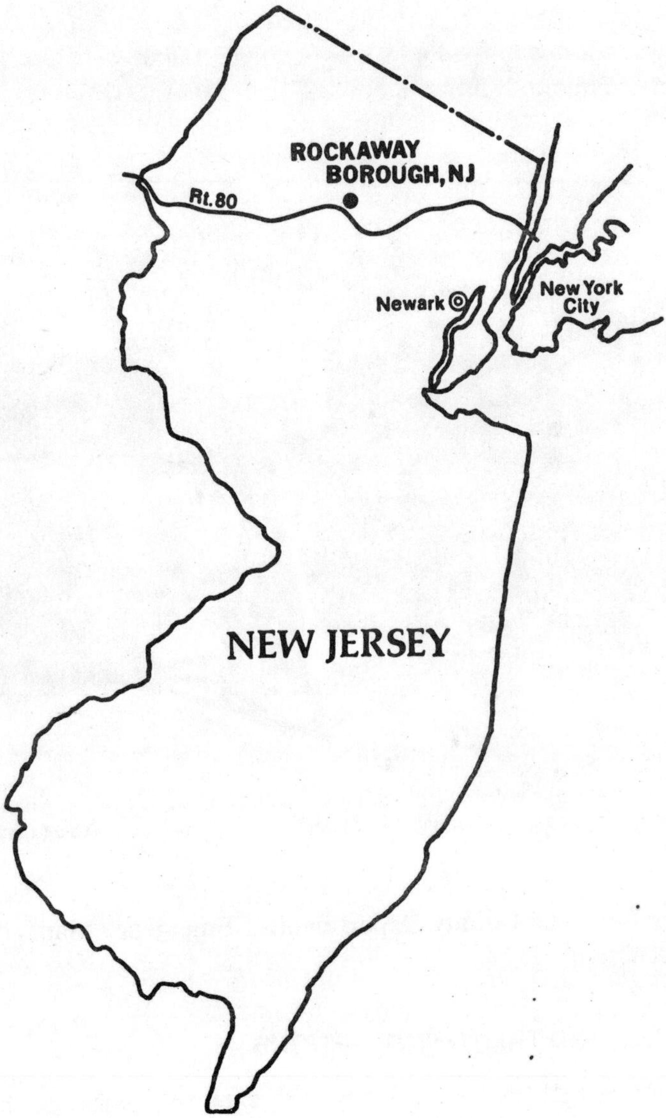


Figure 6-11. Rockaway Borough is located in Morris County in northern New Jersey, *(Courtesy of Calgon Carbon Corporation.)*

Figure 6–12. During the water emergency, the borough provided its residents with free, safe water from Civil Defense and National Guard tank trucks. *(Courtesy of Calgon Carbon Corporation.)*

in Figure 6–12, the County Department of Emergency Management, and a local dairy.

## NEW WELL AND TREATMENT OPTIONS

As already noted, the borough began to explore possible permanent treatment options even before the water emergency was declared.

Dormant municipal wells did not have enough output to supply the borough's 1.5 million gal/day average requirement. Borough land was previously tested for new wells, "but these areas are not sufficient sources of water," according to Krich. "The small excess capacity of nearby municipal systems, combined with 1980–81 drought restrictions on water use, preclude consideration of buying water on a permanent basis from another municipality," Krich said (2).

Aware of the success of carbon treatment at Rockaway Township, the borough contacted Calgon Carbon Corporation to evaluate the use of granular activated-carbon adsorption to treat the groundwater.

Due to the low level of contamination, Calgon was only able to predict performance based upon carbon isotherm information (described in more detail in Chapter 3). A dynamic column study, if properly designed and run, would have taken many months to complete.

Supplier adsorption isotherm data, together with additional published data, was examined. Figure 6–13 shows a carbon adsorption isotherm for tetrachloroethylene published by Dobbs and Cohen of the EPA's Municipal Environmental Research Laboratory in Cincinnati, Ohio (3). This particular isotherm indicates a granular activated-carbon capacity of approximately 35 mg of tetrachloroethylene per gram of carbon at a tetrachloroethylene concentration of 550 ppb.

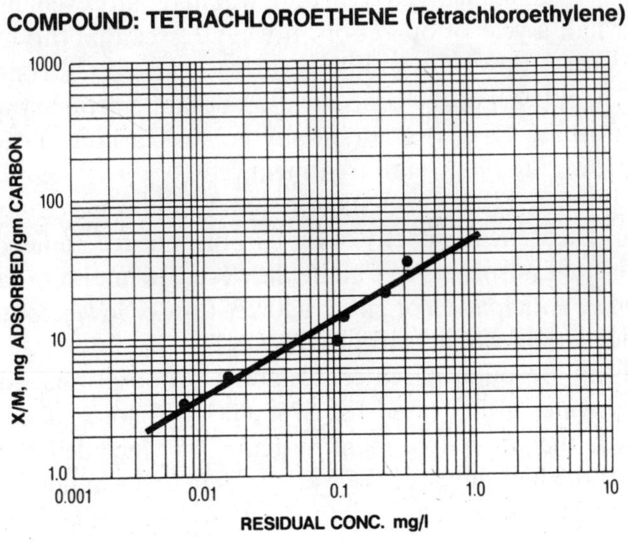


Figure 6–13. Carbon adsorption isotherm plot: tetrachloroethylene. *(Courtesy of Calgon Carbon Corporation.)*

Based upon the favorable isotherm data for adsorption of tetrachloroethylene, installation of three single-stage carbon adsorbers was recommended. Each adsorber would contain 20,000 lb of granular activated carbon, and provide a superficial contact time of 15 min, treating one-third of the total flow. The 15-min contact time has been proven to be sufficient for single-contaminant adsorption at low (ppb) concentrations.

The single-stage adsorber system, which was recommended to minimize capital expenditure, would not allow the maximum use of the granular carbon as predicted by the isotherm results. Calgon estimated that the single-stage capacity might be 60% of the theoretical capacity, or 21 mg/g.

At an influent concentration of 550 ppb, carbon usage was calculated as pounds of carbon per thousand gallons treated:

$$1000\text{ gal} \times 0.55\text{ ppm} \times \frac{8.34\text{ lb PCE}}{\text{ppm }10^6\text{ gal}} \times \frac{\text{lb. carbon}}{0.021\text{ lb. PCE}} = 0.22\text{ lb carbon}$$

Based upon this usage rate, the recommended system, containing 60,000 lb of granular activated carbon, would require exchange of carbon after half a year of operation. It was stressed that this usage rate was valid for a flow of 1.5 million gal/day and a PCE contaminant level of 550 ppb. Once the concentration began to drop in the aquifer (due to flushing effects) substantially greater carbon life would be realized, lowering the overall treatment cost.

It was predicted that a two-stage system would come close to utilizing the full theoretical capacity of the carbon. The recommendation to use single-stage adsorbers was made, however, to minimize capital expense and in anticipation of either a lower flow or lower contaminant level which would extend carbon life.

Rockaway Borough accepted Calgon's proposal and adopted a municipal water utility bond for $700,000 to finance all costs associated with the water treatment system. This included a complete upgrade of the hydraulics of the water system and a building to house the adsorption system. The adsorption system shown in Figures 6–14 and 6–15—consisting of three adsorption vessels, 10 ft in diameter by 20 ft high; vessel internals (collection system); and 60,000 lb of Calgon

14. Within six weeks, Calgon Carbon Corporation delivered three vessels which would comprise the borough's carbon adsorption system. *(Courtesy of Calgon Carbon Corporation.)*

Filtrasorb 300 granular activated carbon—was purchased from Calgon for $136,000. The borough also signed a three-year maintenance agreement with Calgon which would include future carbon deliveries as required.

The adsorption system components were shipped to the borough within six weeks, and the system began treating borough groundwater on June 25, 1981.

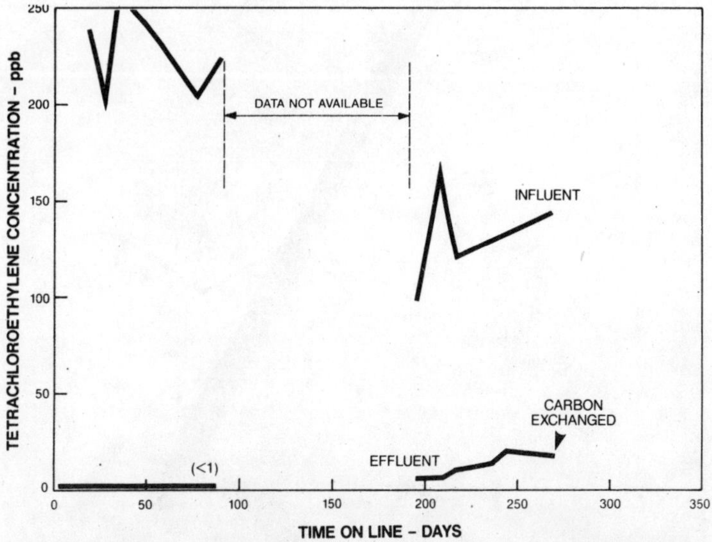


Figure 6–15. Operating results: June 1981 to April 1982. *(Courtesy of Calgon Carbon Corporation.)*

## OPERATING RESULTS

With the initiation of carbon adsorption, the treated water showed no trace of TCE or PCE at the detection limit of 1 ppb and was declared safe for consumption, ending the water emergency.

Weekly testing has continued at the borough, to insure that the water is safe and to identify when the carbon system requires fresh carbon. After flow at all well sites was reinitiated, PCE was detected at levels up to 335 ppb, and TCE at 38 ppb, as influent to the carbon system.

The lower levels of PCE after normal water flow from the wells, and lower average water usage of 1.1 mgd, resulted in longer carbon life than predicted. The initial carbon bed remained on line until April 10, 1982, lasting 288 days. The second carbon bed then treated water until February 15, 1983, lasting 310 days.

Operating results for these two periods are shown in Figures 6–16 and 6–17. These figures illustrate the ability of granular activated carbon to treat a variable influent and provide a safe, consistent effluent,

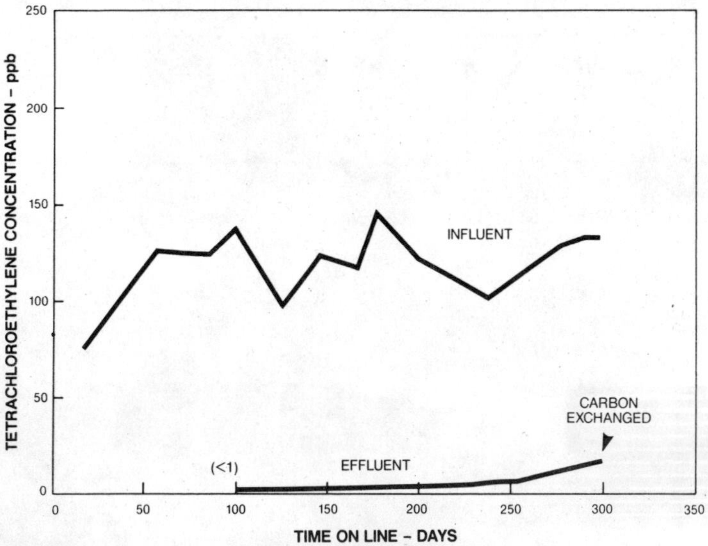


Figure 6–16. Operating results: April 1982 to March 1983. *(Courtesy of Calgon Carbon Corporation.)*

while allowing the user to make a decision when to exchange the carbon.

For the period shown in Figure 6–16, the carbon system treated 362.8 million gal, resulting in a carbon usage rate of 0.165 lb of carbon/1000 gal. It should be noted that this usage was in a single-stage adsorption system, reducing PCE from 130 ppb PCE (avg.) to nondetectable levels. Fresh carbon produced nondetectable levels for about 10 months; when PCE effluent reached 20 ppb the carbon would be replaced.

## ECONOMICS

It is estimated that the borough spent $93,000 in the initial 12 months of water system operation for the carbon treatment system. This expenditure includes amortization of the adsorption system, the maintenance agreement (including additional carbon), and estimated utility costs. With this expenditure, the borough treated 410 million gal at an approximate cost of $0.227/1000 gal.

Figure 6–17. The carbon adsorption system is designed to treat 1.5 million gallons per day in three single-stage adsorbers. The system handles influent with tetrachloroethylene levels as high as 335 ppb. *(Courtesy of Calgon Carbon Corporation.)*

## CONCLUSION

It can be seen from the Rockaway Borough experience that granular activated-carbon treatment of a contaminated groundwater source is a technically and economically feasible solution. The isotherm evaluation illustrates that the technology can be predicted, and the operating experience proves that it can accommodate variable flow and contaminant levels.

The attractiveness of carbon adsorption is best summed up by Mayor Robert Johnson: "How expensive was our solution: to finance the entire project of adsorbers and water system modernization, we raised our water rates $3.00 a month. Our minimum water bill went from \$5.00 per quarter to \$15. The average water bill in the Borough now is between \$60 and \$80 a year." This translates into approximately \$0.76 per person per month.

"In the long run, we feel we have saved our community considerable money by walking the extra financial mile and installing a carbon adsorption system that makes our water pure, guaranteeing the good health of our citizenry," Mayor Johnson said (2).

# BIOLOGICAL TREATMENT OF A GROUNDWATER CONTAMINATED WITH PHENOL

Evan K. Nyer

*DETOX, Inc.*
*Dayton, Ohio*

## INTRODUCTION

A portion of the groundwater under a Gulf Coast hazardous waste site has been contaminated over the course of several years. From analysis of the groundwater and history of the site, several sources have contributed to the problem. These sources have been removed, leaving the cleanup of the groundwater the only remaining task to perform. However, this is the most difficult task, even though the plume is confined to the site boundary.

More than 30 wells have been constructed, and the extent and nature of the contamination plume has been well documented. A plan was developed that would take the water from several central wells, treat it, and then return the treated water through specially constructed recharge pits, forcing the plume back to the central wells. This section discusses the treatment of the groundwater to recharge quality.

The combined central wells have an average concentration of 15,000 mg/liter of dissolved solids and 1300 mg/liter of total organic carbon (TOC) with the main component being 400 mg/liter of phenol. For reinjection, this influent must be treated to background quality, which require a final effluent having less than 18 mg/liter of TOC. The groundwater is naturally brine, so the dissolved solids will not be removed. It was determined that the optimum pumping rate from the wells would be 23,000 gals/day. At this rate of removal and recharge, the design life of the treatment system was set at 10 years.

The process selection for the treatment system had to consider several important and unique problems connected with the treatment of a brine groundwater. Economics of various unit operations were used to determine the proper mix of technologies to reduce the concentration of organics down to the low levels required for recharge. The most critical problem with the design of the treatment system was that the concentration of organics in the groundwater would decrease as the treated water was returned to the ground and forced back to the central wells.

Before the full-scale system was put into service, it was decided to run a large-scale pilot plant to insure that the assumptions made from the laboratory data were correct. The data from the first four months of operation are presented. Certain problems were encountered during startup, but overall, the system performed as expected.

## PROCESS SELECTION

Several factors had to be considered in the process selection for the treatment system. The most important of these were economics and technical factors. A further consideration had to be given to the fact that the site preferred a relatively simple system that would not require full-time monitoring.

Laboratory data showed that the organics in the groundwater could be removed by carbon adsorption or degraded by biological treatment. An economic comparison was then run on the two processes.

The cost of carbon adsorption is directly related to the pounds of organic removed. It takes between 5 and 200 lb of carbon for each pound of organic removed. For high concentrations of organics, as in this groundwater, the range is usually 5–20 lb of carbon per pound of organic. Using 23,000 gal/day and 1300 mg/liter of TOC, 249 lb/day of organics were needed to be removed from the groundwater. Using 10 lb/lb of organic for comparison, 2500 lb of carbon per day were needed. Assuming $0.75/lb of carbon, the operating cost is $1875.00/day. This number does not include capital costs and other operating costs such as personnel and electricity.

Biological treatment is usually considered as cost per 1000 gal of water treated. Standard numbers are based on relatively low concentrations of organics and high flow rates. For small flows (less than 100,000 gal/day) and high organic concentrations, a reasonable cost is $0.46/lb of organic treated. This number is high when compared to textbook numbers, but it is accurate for small-scale systems.

Using the same design numbers of 23,000 gal/day and 1300 mg/liter TOC, the operating cost is $115.00/day. This figure includes capital and electrical costs, but not personnel. This means that carbon will cost at least 16 times the cost of biological treatment.

These cost figures are based on the initial design specifications. The flow through the system will not change during the design life of the project. However, the organic concentration will decrease with time. The groundwater will be removed, treated, and returned to the aquifer. The treated recharge water will be used to force the plume back to the central wells. The treated water will mix with the plume as it forces it forward. This mixing will lower the organic concentration of the groundwater that is being removed. Figure 6–18 shows the expected decrease in organic concentration during the life of the project. These numbers will have to be confirmed after the treatment system is in full operation.

With the change in organic concentration, it was necessary to compare the costs of biological treatment and carbon adsorption over a range of organic concentrations. Figure 6–19 summarizes this comparison. As can be seen in Figure 6–19, the relative cost advantage of biological treatment over carbon adsorption does not change until approximately 150 mg/liter of organic concentration. The costs do not come close until less than 20 mg/liter. From Figure 6–18, the organic concentration is not expected to reach 150 mg/liter until six years after the start of the project.

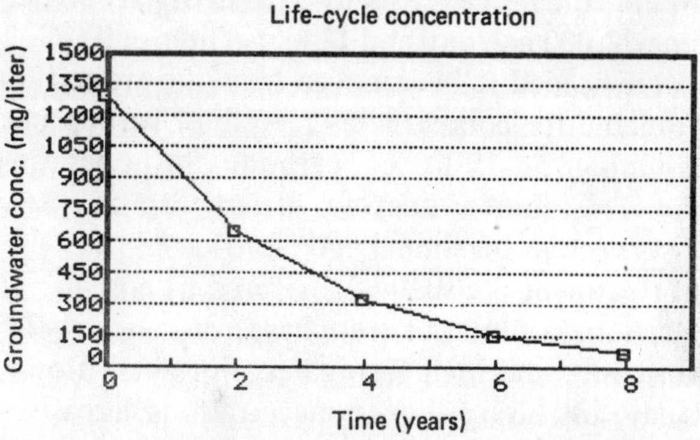


Figure 6-18. Expected decrease in organic concentration.

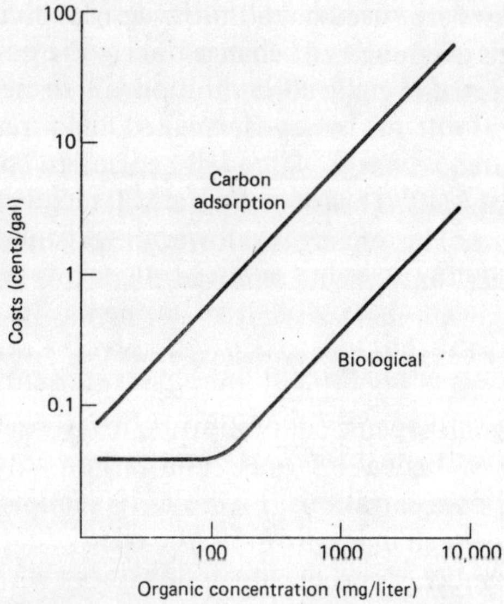


Figure 6-19. Comparison of costs of biological treatment and carbon adsorption.

From the economic analysis, it was determined that biological treatment of the groundwater was the preferred method. The only problem was that the treatment facility would have to consistently produce 18 mg/liter TOC for the recharge water. Based on experience and various literature on biological treatment, doubts were expressed that the biological system could consistently produce 18 mg/liter.

The solution was to use biological treatment followed by carbon adsorption. This conclusion was arrived at without the use of a pilot plant study, because it was determined that the capital cost of a carbon system would be less than the cost of a pilot plant to test whether a carbon system was necessary.

## LIFE-CYCLE DESIGN

There are several designs for biological treatment systems. The easiest to run are the fixed-film designs. However, the activated-sludge designs produce the best effluent; furthermore, several tanks were available on site that could be used as part of the system, and an activated-sludge design could use the existing tanks.

Activated sludge was, therefore, preferable on a capital cost basis. However, there were two technical problems with using activated sludge. First, in order to maintain a consistent, low effluent organic level, the activated-sludge process requires close operator attention and daily analytical support. As mentioned before, these personnel requirements add substantial costs to the operation of the system. It was estimated that an activated-sludge design would require at least eight man-hours per day.

The second problem was that an activated-sludge system would not be able to adjust to the lower organic concentrations during the design life of the project. Figure 6–20 summarizes the effect of lower influent organic levels on the sludge age of the treatment system. This analysis assumes 23,000 gal/day influent flow, 40,000-gal aeration tank, and 0.25 lb solids/lb TOC removed yield coefficient.

The activated-sludge process relies on the ability of the bacteria to settle. A sludge age between 5 and 20 days is recommended to maintain a good settling sludge. As can be seen in Figure 6–20, the sludge age quickly goes out of that range as the influent concentration goes

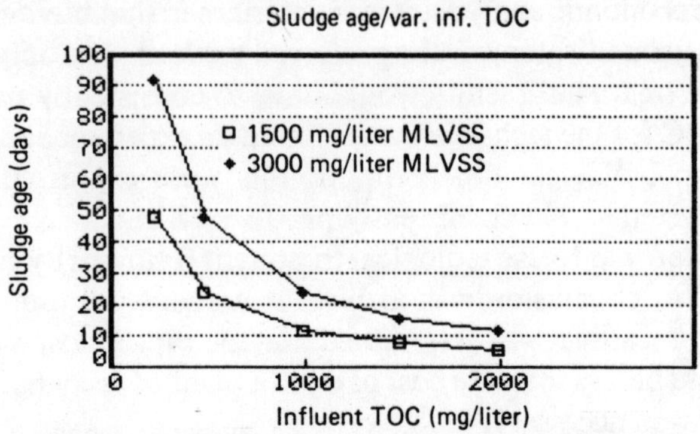


Figure 6–20. Effect of lower influent organic levels on sludge age.

down. In the beginning of the project, a large aeration tank was required, and as the influent concentration goes down, the aeration basin must shrink to maintain the proper sludge age.

All of these considerations were combined and a final design developed. Figure 6–21 is the final design for the groundwater treatment system. The system includes a first-stage activated-sludge system, a second-stage fixed-film/activated-sludge system, a dual-media filter, and a carbon adsorption column. The following are the specifications on each section:

First-Stage Biological—The system consists of two 20,000-gal (15-ft diameter, 15-ft height) aeration basins, in series, and a hopper bottom clarifier. The first aeration basin has eight static tube aerators and a 15-hp blower delivering 240 scfm of air. The second aeration basin has four static tube aerators and a 5-hp blower delivering 80 scfm of air. The hopper bottom clarifier has 97 $ft^2$ of surface area and returns sludge with an air, sludge ejector pump.

Second-Stage Biological—The second stage is the Fixed Activated Sludge Treatment System or FAST (registered trademark of Smith & Loveless, Inc.). The unit is 10 ft wide, 10 ft high, and 28 ft long. The system maintains the bacteria in the aeration zone by attachment to plastic media. The media are submerged in the water and the tank is completely mixed. The resulting system is an activated-sludge system,

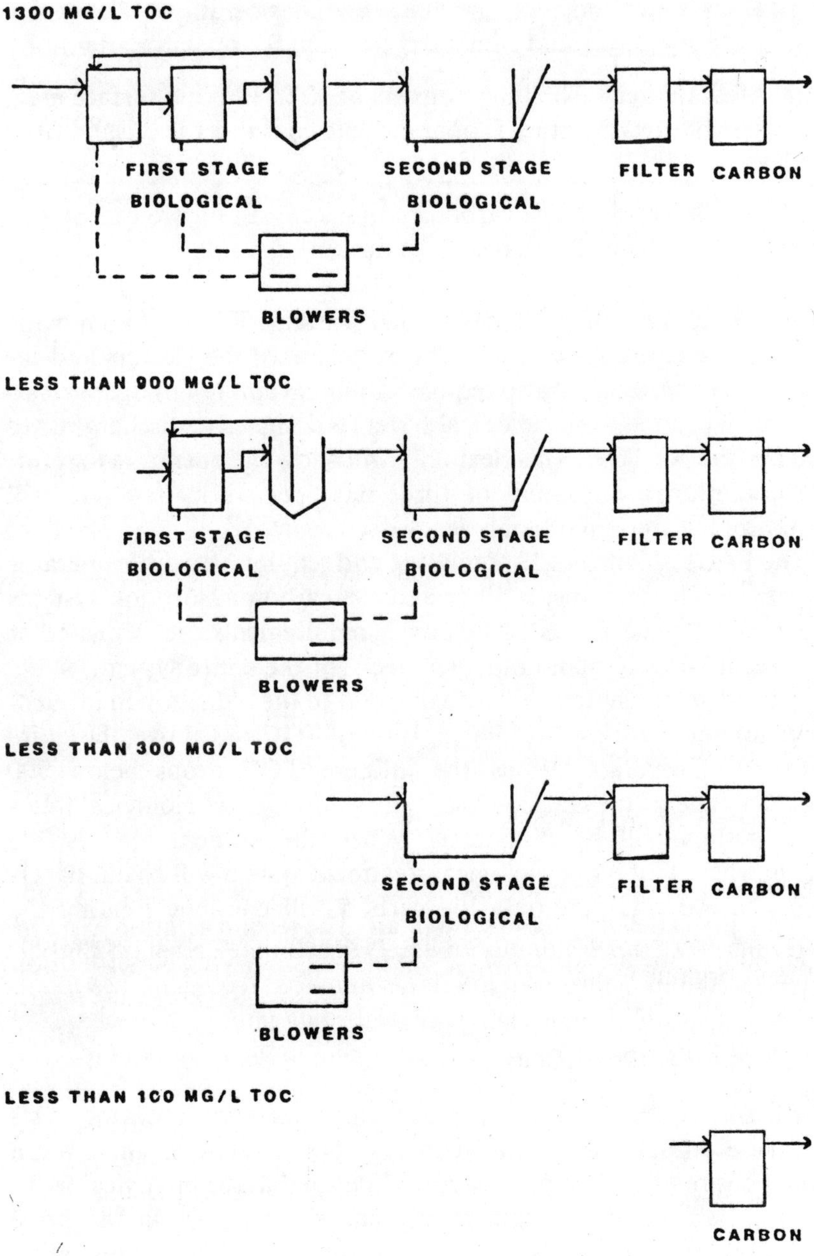


Figure 6-21. Groundwater treatment schemes at various influent organic levels.

but it is no longer limited by the sludge age considerations. The system is also self-regulating and eliminates the need for operator attention.

Dual-Media Filter—The filter consists of 10 $ft^2$ filtering surface area, air / water backwash with 1 ft of anthracite coal and 1 ft of sand filter media.

Carbon Adsorption—The carbon columns consist of two carbon columns in series with 30 min residence time in each column.

The resulting system will be easy to maintain. There are no moving parts in the entire system, with the exception of the blowers and the pumps. The system will also require a minimal amount of operator attention. The first-stage biological system is designed to discharge up to 300 mg / liter of TOC. This flexibility will allow the operators to refine the operation a maximum of three days per week, and also will minimize the amount of analytical work required.

The FAST system is self-regulating and requires very little operator attention. The dual-media filter and the carbon adsorption systems are both fully automated. Manpower requirements are estimated at between 12 and 20 man-hours per week for the entire system.

The treatment system will also respond to the reduction in influent organic concentration. At 1300 mg / liter of TOC and above, the entire system will operate. When the influent TOC drops below 900 mg / liter, one of the aeration basins in the first-stage biological treatment system will be eliminated. When the influent reaches 300 mg / liter, the first-stage biological treatment system will go out of service. At 100 mg / liter, only the carbon will continue treating the groundwater. Figure 6–21 shows the treatment schemes at the various influent organic levels.

## PILOT PLANT OPERATION

The full-scale plant design was based on preliminary laboratory tests and the economic and technical analysis presented here. There was a strong desire to have more concrete evidence that the biological treatment system would be able to degrade the organics in the brine groundwater. However, it was estimated that a continuous operating pilot plant would cost on the order of 25% of a full-scale system. It must be remembered that the full-scale system is only 23,000 gal/day.

The decision was to set up a pilot plant and have all of the material used on the pilot plant capable of being used on the full-scale system. The hopper bottom clarifier was purchased, and the second aeration tank of the first-stage biological system was set up with four static tube aerators and the 5-hp, 80-scfm blower. Both of these units would be used on the full-scale system. The pilot plant was estimated to be capable of treating 5000 gal/day.

On June 17, 1983, the pilot plant was started up. The initial seed for the reactor was provided by 8000 gal of waste-activated sludge from a nearby refinery and a commercial bacteria culture specific for hydrocarbons in salt water. The initial response of the bacteria was immediate. The dissolved oxygen in the aeration basin quickly went to less than 1 mg/liter. The influent flow was, at first, controlled by the level of dissolved oxygen in the tank.

The initial concentration of salt in the aeration tank was low. As brine groundwater was fed, the concentration increased. The salt had a definite effect on the bacteria. The mixed liquor suspended solids (MLSS) never increased past the initial concentration, and by day 14, started to decrease. Organic removal was on the order of 60% but the bacterial population would not increase.

At this point, the operator also switched methods for determining phosphorus and ammonia in the brine water. In both cases, the concentrations were found to be below desired levels. On day 26, 5000 gal of waste-activated sludge from a refinery that treated ballast from oil tankers was added. These bacteria had grown in high concentrations of salt. At the same time, nutrients were added on a daily basis.

The bacterial population took about eight days to respond. After that, the MLSS continuously increased in concentration. On day 68, the MLSS reached 6000 mg/liter, and bacterial solids were wasted from the system. MLSS was maintained between 6000 and 7000 mg/liter. Figure 6–22 summarizes the MLSS levels in the pilot plant.

Effluent TOC responded to the increase in MLSS. Figure 6–23 summarizes the TOC concentration in the effluent for the pilot plant. On day 88, the influent flow increased to 15,000 gal/day. As can be seen in Figure 6–23, the effluent TOC increased with the added load. The

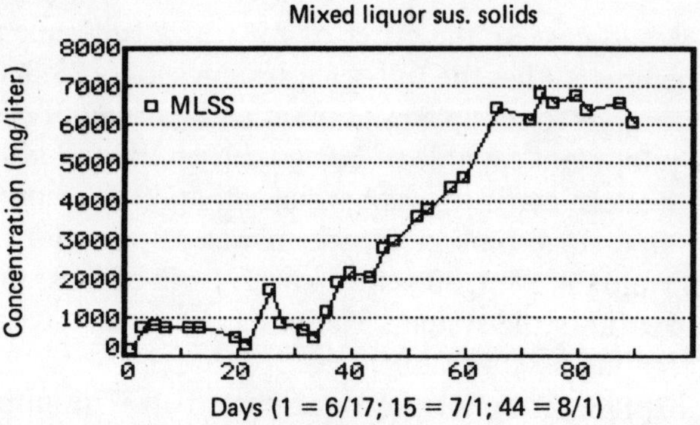


Figure 6-22. MLSS levels in the pilot plant.

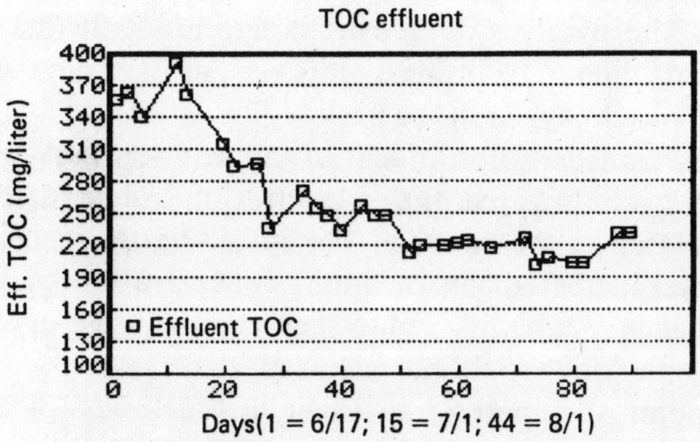


Figure 6-23. TOC concentration in the effluent for the pilot plant.

cause of the increase was a faulty valve controlling the flow to the system.

Figure 6-24 summarizes the TOC removed by the pilot plant. In general, the TOC removed follows the same pattern as the MLSS and the effluent TOC. However, the same valve problem has made it difficult to control the flow to the pilot plant. The TOC removed has had a lot of variation due to the uncontrolled influent flow. The full-scale system will have a better control method for influent flow.

As can be seen by the above data, the organics in the brine groundwater are degradable by biological treatment. The entire treatment system was, therefore, installed.

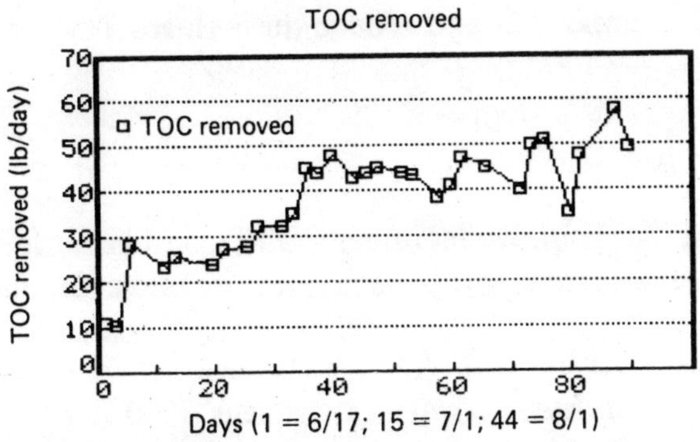

Figure 6-24. TOC removed by the pilot plant.

## FULL-SCALE OPERATION

The full-scale plant could not be put into operation at one time. The carbon unit could not be put into operation before the second-stage biological treatment system was operating. The dual-media filter was not required until the carbon unit was operational. The first step toward full-scale operation was to bring the second tank of the first-stage biological unit on line. The second step was to install the FAST unit.

The first-stage biological unit was able to have an immediate startup. This was due to the established bacteria from the pilot operation. The FAST system took the more normal time of six weeks to start up. During this startup time, the effluent from the biological treatment systems was sent to an evaporation pond. Flow to the system was maintained at below full design flow to minimize the load on the pond.

Once the FAST system was performing, the dual-media filter and the carbon adsorption units were put into operation. Suspended solids from the FAST unit were very low, and the dual-media filter was put on a timed backwash cycle. The carbon was able to consistently remove the remaining TOC to 18 mg/liter. The influent to the carbon system was a little higher than expected. The average influent to the carbon was 120 mg/liter. A project has now been started to seed the biological reactors with bacteria that can degrade some of the refractory organic compounds and reduce the influent TOC to the carbon units.

## CONCLUSIONS

A portion of the groundwater under a Gulf Coast hazardous waste site has been contaminated with a variety of organic compounds. It was decided to pump the groundwater out of the ground, remove the organic contaminants, and recharge the cleaned water back into the ground. Initial influent concentration was 1300 mg/liter of TOC, flowing at 23,000 gal/day. Recharge concentration was set at the background concentration of the groundwater, 18 mg/liter of TOC. Over the life of the project, the influent concentration will approach the background concentration.

A treatment scheme was developed based on laboratory tests, and technical and economic analysis. The final system included the following: a first-stage activated-sludge system, a second-stage fixed-film/activated-sludge system, a dual-media filter, and a carbon adsorption column. This system was economical and could easily be changed to reflect the changes in the influent concentration.

To insure that the organics in the brine groundwater could be degraded by bacteria, a pilot plant was set up. The pilot plant was designed so that all of the components would be used on the full-scale system. After overcoming salt inhibition and low nutrient concentrations, the pilot plant was able to consistently remove 70% of the TOC. The full-scale plant was started up in the first quarter of 1984, and it has steadily produced effluent TOC levels of less than 18 mg/liter.

# IN-SITU BIOLOGICAL TREATMENT OF ISOPROPANOL, ACETONE, AND TETRAHYDROFURAN IN THE SOIL/GROUNDWATER ENVIRONMENT *

Paul E. Flathman
and
Gregory D. Githens

*O. H. Materials Co.*
*Findlay, Ohio*

## INTRODUCTION

In-situ biological cleanup following spills of biodegradable hazardous organic compounds in the soil/groundwater environment can be a cost-effective technique when proper engineering controls are applied (4, 5, 6). Biodegradation of hazardous organic contaminants by microorganisms (7) can minimize liability by converting toxic reactants into harmless end products.

The cleanup of soil and groundwater containing isopropanol (IPA), acetone, and tetrahydrofuran (THF) by a combination of biological and physical techniques resulted in 90% removal of IPA and THF within three weeks. Acetone, an intermediate oxidation product of IPA metabolism, was removed by the end of the sixth week. The spill originated from several buried tanks which leaked contents into a shallow basin (12 ft) containing 100,000 ft$^3$ of sand and pea gravel.

The initial problem was to determine if biological cleanup could be utilized to remove isopropanol and acetone; 500 gal of THF were spilled by the client at an early stage of field operation. Acetone had not been spilled at the site but was suspected to be an incomplete oxidation product of isopropanol metabolisms (8).

*This section provided courtesy of O. H. Materials Co.

This case history describes

1. Bench-scale evaluation of the potential for biological cleanup in the spill site matrix
2. Field implementation
3. Removal rates of the contaminants at the spill site

Since the client is confidential, additional background information regarding this project cannot be presented. The underground recovery and treatment system, designed and developed by O. H. Materials Co., was used to effectively remove and treat the organic contaminants in the soil/groundwater environment.

O. H. Materials Co. has performed biological cleanups of spilled substances since 1978, when a railroad incident resulted in spillage of acrylonitrile. Subsequent biological environmental restoration projects have included additional acrylonitrile spills and other materials such as gasoline, crude oil, ethylene glycol, butylcellosolve, ethylacrylate, n-butylacrylate, methylene chloride, and various phenolics.

The case history demonstrates the practicality of biological detoxification of certain contaminants in the soil/groundwater environment. The data presented support O. H. Materials Co.'s earlier work and findings on groundwater restoration using biological techniques. Using the underground recovery and treatment system in combination with an activated-sludge biological treatment system, a recent project (6) also achieved greater than 50% reduction in the groundwater concentration of ethylene glycol within one week.

For this project, a recovery system was used to withdraw contaminated water from the ground for aboveground biological treatment. Effluent from the treatment system was reinjected into the subsurface environment, creating a closed-loop system. Biodegradation of the contaminants took place in the soil/groundwater environment as well as aboveground in the biological treatment system. The injection system was used to inoculate the underground environment with microbes capable of biodegrading the organic contaminants and to provide the inorganic nitrogen and phosphorus necessary to support microbial growth.

Cost effectiveness, minimal disturbance to existing operations, on-site destruction of spilled compounds, and permanence of solution are several of the advantages identified in this project for implementing biodegradation as a technique for spill cleanup and environmental restoration.

## BIOFEASIBILITY EVALUATION

Prior to field operation, data were analyzed and a laboratory study was performed to determine whether a biodegradation program at the site was feasible. Results of analyses performed for soil and groundwater samples collected from the contaminated area and from a clean control area indicated that the value of all operational parameters were within favorable ranges for an enhanced biodegradation effort and only nitrogen and phosphorus needed to be augmented.

The biodegradation feasibility study was performed using static flask culture techniques (9, 10). Isopropanol-acetone-basal salts media (500 ml/liter Erlenmeyer flask) were inoculated with 12.5 ml (11) of a 1% (wt/vol) soil suspension and incubated at ambient temperature. The soil used for the inoculum was a representative sample collected from the spill site.

The medium employed (12) was prepared as follows:

| | |
|---|---|
| Isopropanol | 130 mg |
| Acetone | 60 mg |
| $KH_2PO_4$ | 0.4 g |
| $K_2HPO_4$ | 1.6 g |
| $NH_4NO_3$ | 0.5 g |
| $MgSO_4 \bullet 7H_2O$ | 0.2 g |
| $CaCl_2 \bullet 2H_2O$ | 0.025 g |
| $FeCl_3 \bullet 6H_2O$ | 0.0025 g |
| per liter | |

The pH of the medium was adjusted to 7.2 with 1*N* HCl or 1*N* NaOH prior to inoculation. An uninoculated control was prepared to quantify nonbiological loss of isopropanol and acetone.

At periodic intervals throughout the study, aliquots were removed from the reaction mixtures and analyzed for isopropanol and acetone and for pH and $NH_3$-N, $NO_3$-N, and $PO_4$-P concentrations to insure maintenance of a chemical environment favorable for bacterial growth. Isopropanol and acetone were quantified by headspace analysis on a Perkin-Elmer HS-6 unit (Perkin-Elmer Corporation, Instrument Division, Norwalk, Connecticut) in conjunction with a Tracor 560 gas chromatograph (Tracor, Inc., Instrument Group, Austin, Texas). The gas chromatograph was equipped with a 6-ft 1% SP-1000 on a 60/80 mesh Carbopack B column with a flame ionization detector. Ammonium, nitrate, and ortho-phosphate were quantified spectrophotometrically by nesslerization, cadmium reduction, and ascorbic acid, respectively (Hach Chemical Company, Ames, Iowa).

Figure 6–25 indicates that isopropanol was biodegraded stoichiometrically in the reaction vessel to acetone. This reaction was biological since it did not take place in the uninoculated control. Daily biological loss during the first eight days of the feasibility study was 5.8 mg/liter, and nonbiological loss was only 0.39 mg/liter. Thus, the rate of biological loss for isopropanol was 15 times greater.

The concentration of acetone continued to increase until isopropanol was biodegraded to a trace amount and rapidly decreased once isopropanol was completely metabolized. The lack of a significant lag period indicated the presence of adapted indigenous microbes which could biodegrade the contaminants. This finding was significant because it indicated that in-situ biodegradation of isopropanol was already occurring and that OHM's management approach should be to increase the natural biodegradation rate. A biodegradation feasibility study was not performed for THF since it was not an original contaminant. THF, however, has been shown to be biodegradable by adapted microorganisms (13).

## FIELD IMPLEMENTATION

Use of the underground recovery and treatment system for aquifer restoration has been previously described (5, 6, 14, 15, 16). All components of the system are easily assembled on site so that startup can take place within a matter of hours. With cleanup completed, those

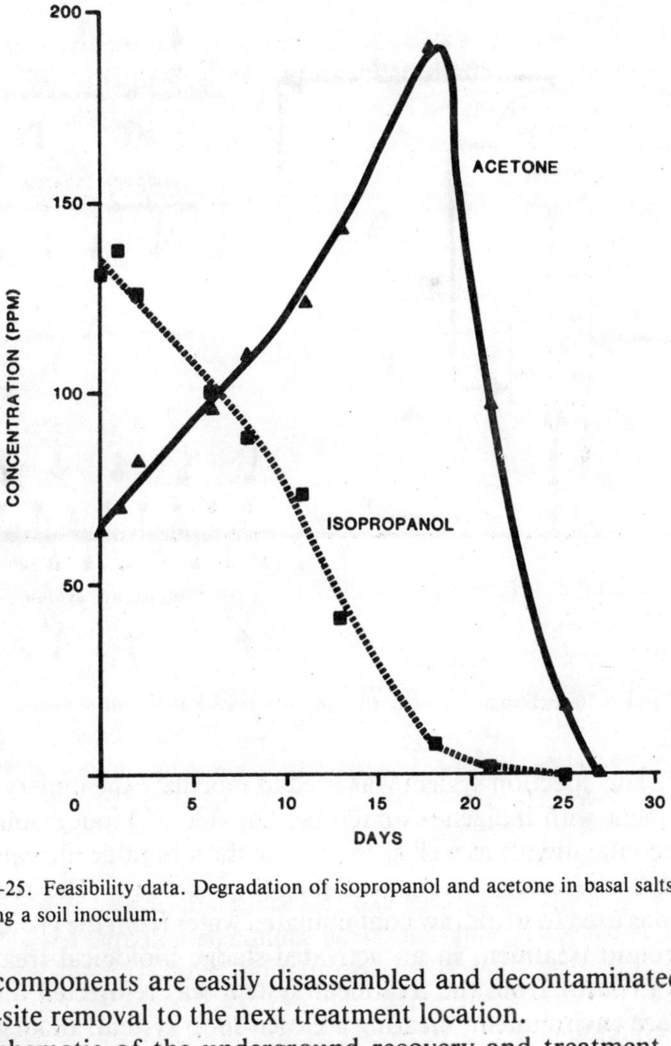


Figure 6–25. Feasibility data. Degradation of isopropanol and acetone in basal salts solution containing a soil inoculum.

same components are easily disassembled and decontaminated prior to off-site removal to the next treatment location.

A schematic of the underground recovery and treatment system used for this project is presented in Figure 6–26. An activated-sludge system was the preferred method for aboveground biological treatment. In addition to providing efficient biological treatment, the activated-sludge system permitted wasting of adapted microorganisms for inoculation of the soil/groundwater environment through the injection system. With a flow rate of 6–10 gpm; the treatment system was designed to provide a triple flush of the spill area within two

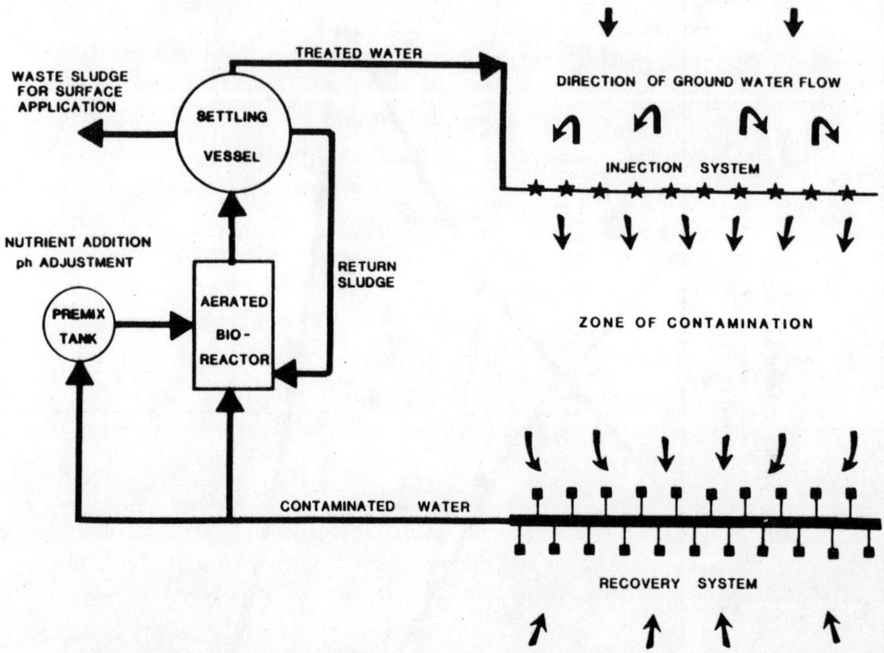


Figure 6-26. Schematic of the underground recovery and treatment system.

months. The injection system was used to inoculate the underground environment with indigenous microbes capable of biodegrading the organic contaminants as well as to provide the inorganic nitrogen and phosphorus necessary to support microbial growth. The recovery system was used to withdraw contaminated water from the ground for aboveground treatment in an activated-sludge biological treatment system. Effluent from the treatment system was reinjected into the subsurface environment, creating a closed-loop system. Biodegradation of the spilled organic contaminants took place in the subsurface soil/groundwater environment as well as aboveground in a biological treatment system.

Nitrogen and phosphorus nutrient additions and pH adjustments were made to the recovered groundwater from the recovery wells through the use of a premix tank to the bioreactor. From the bioreactor, the treated groundwater was pumped into a settling chamber from which a portion of the settled sludge was recycled into the bioreactor. Excess sludge was periodically wasted from the treatment system. The remainder of the treated groundwater was used for the injection system.

Results in Figures 6–27, 6–28, and 6–29 present respective IPA, acetone, and THF concentrations over a 35-day period for recovery system effluent and for a centrally located test well. Patterns of con-

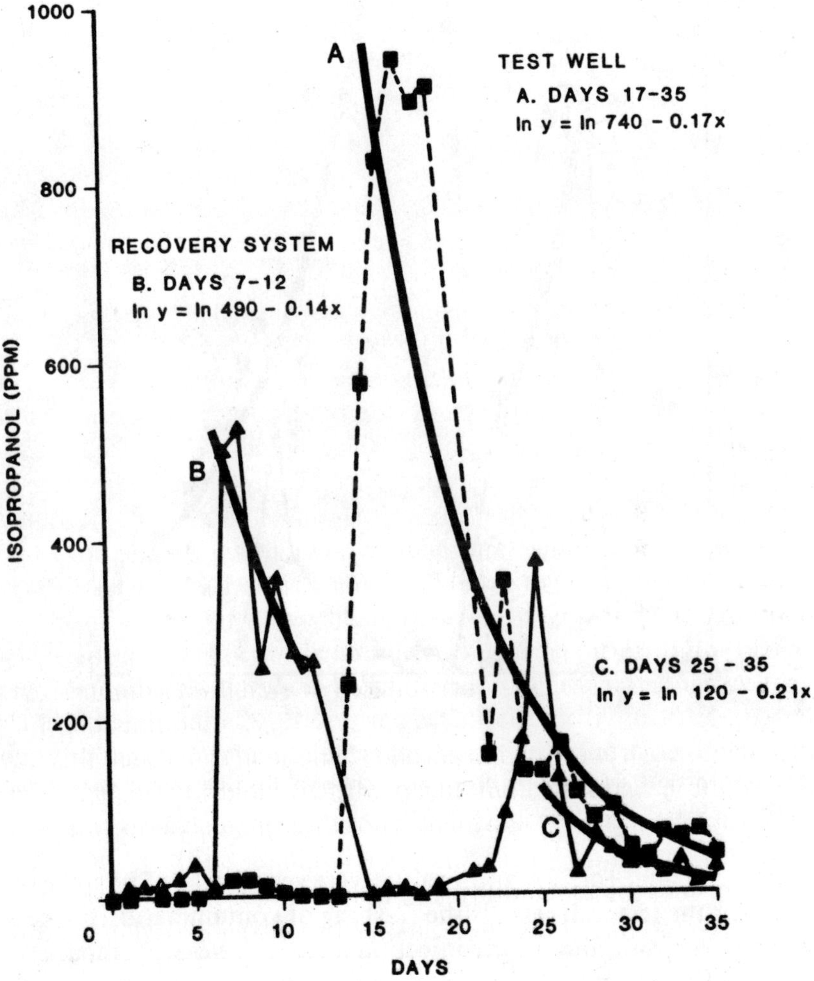


Figure 6–27. Isopropanol field data. IPA concentrations in the underground recovery and treatment system.

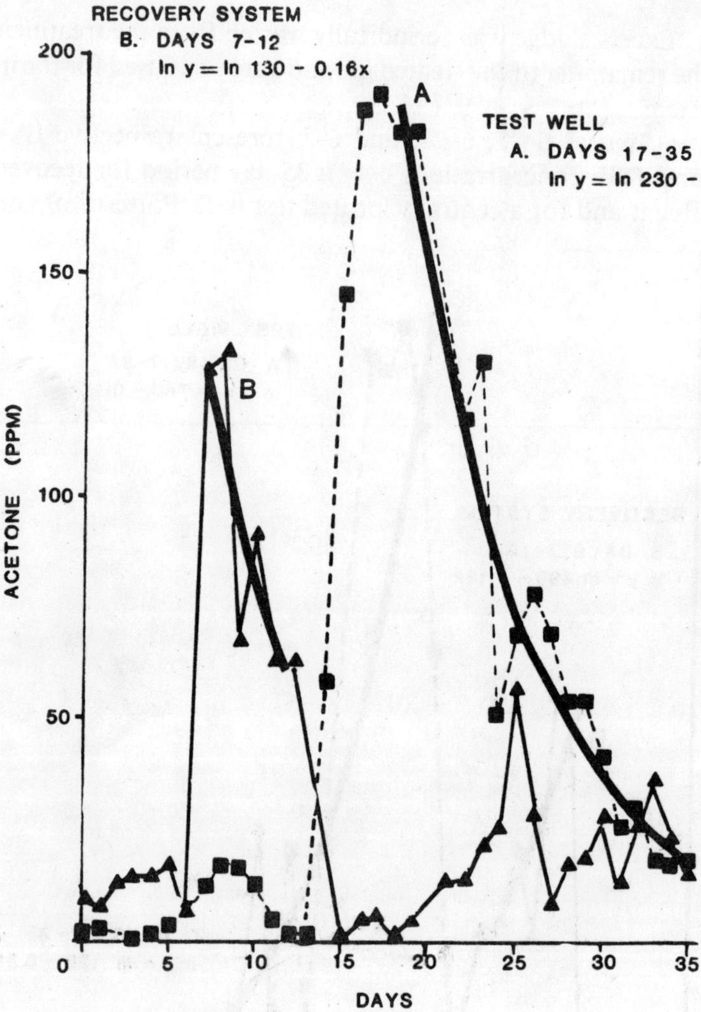


Figure 6-28. Acetone field data. Acetone concentrations in the underground recovery and treatment system.

centration change for IPA and acetone were very similar. The spike on day 15 for the test well reflects the flushing of contamination pockets following repositioning of the injection/recovery wells. Comparison of Figures 6-27 and 6-28 supports initial bench-scale findings of biological conversion of IPA to acetone After day 15, acetone con-

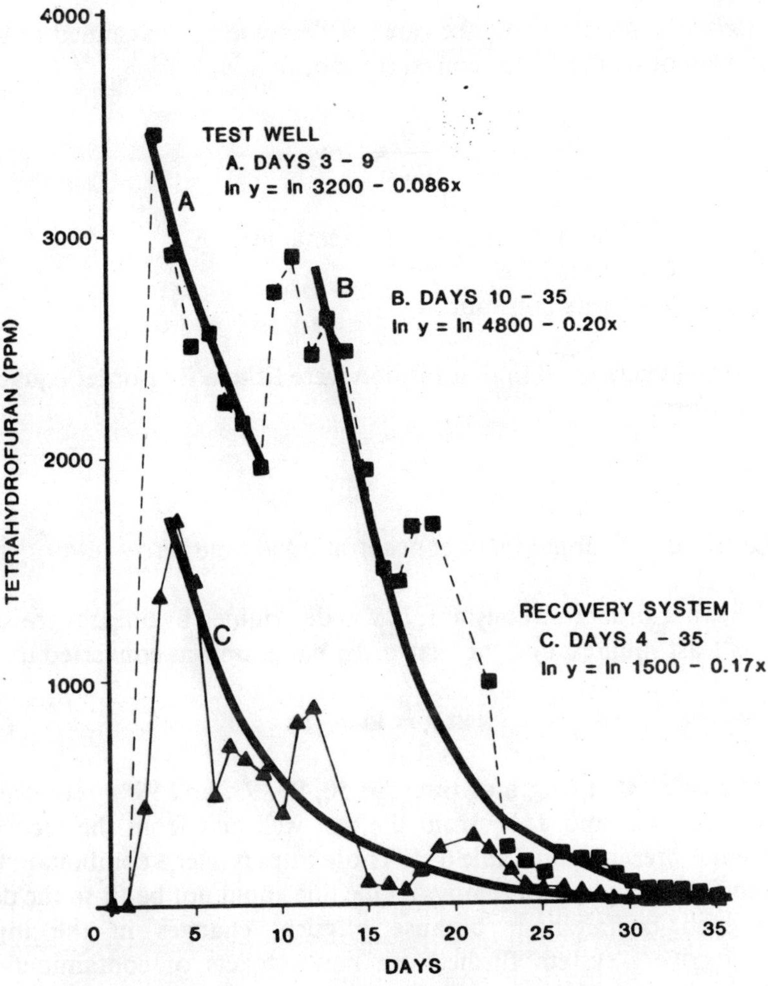


Figure 6-29. Tetrahydrofuran field data. THF concentrations in the underground recovery and treatment system.

centration continued to increase until a low concentration of IPA was reached and then decreased very rapidly. By day 38, acetone concentration was less than 2 mg/liter and by day 44 was below the detection limit of 0.2 mg/liter.

Exponential decay curves were used to quantify removal rates of IPA, acetone, and THF from the groundwater environment. Holding other variables constant, the rates of decrease were assumed to be a function of contaminant concentration, that is,

$$\frac{dy}{dx} = -by$$

where: $y$ = level of contaminant remaining
$x$ = time
$b$ = rate constant

The curves generated in that fashion were fit to a first-order equation of the form

$$y = ae^{-bx}$$

where: $a$ = contaminant concentration at time zero

The first-order rate constant, $b$, was determined by linear regression using least squares, and the first-order equation was converted to

$$\ln y = \ln a - bx$$

The calculated length of time for 50, 90, 95, and 99% removal of IPA, acetone, and THF from the test well and from the recovery system is presented in Table 6–4. Those rates reflect a combination of biological and physical removal. One line could not be fit to the data for each contaminant because physical changes in the injection/recovery system flushed out new pockets of contamination. Comparison of THF test well data following adjustment of the injection/recovery system on day 10 indicated that the calculated removal time was reduced by more than 50%.

## DISCUSSION AND CONCLUSION

The combination of physical and biological techniques was effective in removing those contaminants at a relatively rapid rate. The flexibility of the injection/recovery system in inoculating the site with

**TABLE 6-4. Contaminant Removal Rates Using the Underground Recovery and Treatment System with Biological Techniques.**

| | DAYS | | | | | |
|---|---|---|---|---|---|---|
| | 50% REMOVAL | 90% REMOVAL | 95% REMOVAL | 99% REMOVAL | SAMPLE SIZE $n$ | COEFFICIENT OF DETERMINATION $r^2$ |
| Isopropanol | | | | | | |
| Test well | | | | | | |
| Curve A | 4 | 14 | 18 | 27[a] | 17 | 0.86 |
| Recovery system | | | | | | |
| Curve B | 5 | 16[a] | 21[a] | 33[a] | 6 | 0.63 |
| Curve C | 3 | 11[a] | 14[a] | 22[a] | 11 | 0.55 |
| Acetone | | | | | | |
| Test well | | | | | | |
| Curve A | 5 | 16 | 21[a] | 33[a] | 17 | 0.93 |
| Recovery system | | | | | | |
| Curve B | 4 | 14[a] | 19[a] | 29[a] | 6 | 0.73 |
| Tetrahydrofuran | | | | | | |
| Test well | | | | | | |
| Curve A | 8[a] | 27[a] | 35[a] | 54[a] | 7 | 0.94 |
| Curve B | 4 | 12 | 15 | 23 | 24 | 0.95 |
| Recovery system | | | | | | |
| Curve C | 4 | 13 | 18 | 27 | 29 | 0.86 |

[a]Extrapolated values.

biological and flushing media was a key determinant in the removal of scattered pockets of contamination.

For this project, two shallow basins of similar size and with similar contaminants were successfully treated using biological techniques. If it had been necessary to remove the contaminated soil from the two shallow basins treated at the site, the transportation and disposal (T&D) costs for 200,000 ft$^3$ would have been at least $550,000. The estimate for traditional T&D was based on a transportation cost of $3.00/mi/truck to a hazardous waste disposal facility 100 mi from the treatment site. Twenty cubic yards was the calculated capacity for each truck. The disposal cost for the contaminated soil was $60.00/yd$^3$.

By successfully decontaminating the soil and allowing it to remain on site, more than a five-fold cost savings was achieved and all future liability was substantially reduced. This estimate for cost savings, however, is conservative, since removal of soil from an area increases void volume for transportation and disposal by a factor of 1.2–1.3. Many times, hazardous materials cannot be shipped off site in bulk but must be packaged into drums. Additional chemical analyses are generally required to characterize the waste prior to off-site transportation and disposal. Inclusion of all these additional factors that must be considered for off-site disposal resulted in an even greater cost benefit to the client for in-situ cleanup.

Biodegradation as a method for spill cleanup and environmental restoration is considered a promising technology (5, 6, 17, 18). Land treatment techniques have been engineered and are accepted as an economical and environmentally sound means of destruction for many types of industrial wastes (5, 19, 20, 21). With regard to the cleanup of contaminated soil and groundwater, physical removal, by convention, has been a common method for remediation. However, biological techniques are now gaining increasing acceptance as a practical, cost-effective alternative for environmental restoration.

## REFERENCES

1. Pye, Veronica I., Patrick, Ruth, and Quarles, John. "Groundwater Contamination in the United States." Philadelphia: University of Pennsylvania Press, 1983.
2. When a Water Supply Went Bad. *American City & County* Magazine, December 1981.
3. Dobbs, Richard A., and Cohen, Jesse M. Carbon Adsorption Isotherms for Toxic Organics. Cincinnati, Ohio: U.S. EPA Municipal Environmental Research Laboratory (EPA-600/8-80-023), April 1980.
4. Current Developments, Hazardous Waste. *Environmental Reporter,* **May** 6:11–12 (1983).
5. Flathman, P. E., Studabaker, W. C., Githens, G. D., and Muller, B. W. Biological Spill Cleanup. Proceedings of the Technical Seminar on Chemical Spills, Toronto, Ontario, Canada, October 25–27, 1983. Ottawa, Ontario, Canada: Technical Services Branch, Environmental Protection Service, Environment Canada, pp. 117–130.
6. Flathman, P. E., Quince, J. R., and Bottomley, L. S. Biological Treatment of Ethylene Glycol-Contaminated Groundwater at Naval Air Engineering Center, Lakehurst, New Jersey. Proceedings of the Fourth National Symposium and Exposition on Aquifer Restoration and Ground Water Monitoring, Columbus, Ohio, May 23–25, 1984. Worthington, Ohio: National Water Well Association. In preparation.

7. Kobayashi, H., and Rittmann, B. E. Microbial Removal of Hazardous Organic Compounds. *Environ. Sci. Technol.,* **16**:170A–183A (1982).
8. Stanier, R. Y., Doudoroff, M., and Adelberg, E. A. "The Microbial World," 3rd ed. Englewood Cliffs, N.J.: Prentice-Hall, 1970.
9. Barth, E. F., and Bunch, R. L. Biodegradation and Treatability of Specific Pollutants, EPA-600/9-79-034. Cincinnati, Ohio: Municipal Environmental Research Laboratory, U.S. EPA, 1979.
10. Tabak, H. H., Quave, S. A., Mashni, C. I., and Barth, E. E. Biodegradability Studies with Organic Priority Pollutant Compounds. *J. Water Pollut. Control Fed.,* **53**:1503–1518 (1981).
11. Haller, H. D. Degradation of Mono-Substituted Benzoates and Phenols by Wastewater. *J. Water Pollut. Control Fed.,* **50**:2771–2777 (1978).
12. Horvath, R. S., and Alexander, M. Cometabolism of *m*-Chlorobenzoate by an *Arthrobacter. Appl. Microbiol.,* **20**:254–258 (1970).
13. E. I. du Pont de Nemours & Company, Inc. Tetrahydrofuran—Properties, Uses, Storage & Handling. Wilmington, Del: Publication no. E-62465, 1984, p. 28.
14. Ohneck, R. J., and Gardner, G. L. Restoration of an Aquifer Contaminated by an Accidental Spill of Organic Chemicals. *Ground Water Monitoring Review,* **2**(4):50–53 (1982).
15. Quince, J. R., and Gardner, G. L. Recovery and Treatment of Contaminated Groundwater: Part I. *Ground-Water Monitoring Review,* **2**(3):18–22 (1982).
16. Quince, J. R., and Gardner, G. L. Recovery and Treatment of Contaminated Groundwater: Part II. *Ground Water Monitoring Review,* **2**(4):18–25 (1982).
17. Dobbs, D., and Walton, G. C. Biodegradation of Hazardous Materials in Spill Situations. Paper presented at 1980 National Conference on Control of Hazardous Material Spills. Louisville, Ky., May 13–15, 1980.
18. Thibault, G. T., and Elliott, N. W. Biological Detoxification of Hazardous Organic Chemical Spills. Paper presented at 1980 National Conference on Control of Hazardous Material Spills. Louisville, Ky., May 13–15, 1980.
19. Loehr, R. C., Jewell, W. J., Novak, J. D., Clarkson, W. W., and Friedman, G. S. "Land Application of Wastes" (2 vols.). New York: Van Nostrand Reinhold, 1979.
20. Parr, J. F., Marsh, P. B., and Kla, J. M., "Land Treatment of Hazardous Wastes." Park Ridge, N.J.: Noyes Data Corporation, 1983.
21. Vernick, A. S., and Walker, E. C., eds. "Handbook of Wastewater Treatment Processes." New York: Marcel Dekker, 1981.

were made to the recovered groundwater from the recovery wells through the use of a premix tank to the bioreactor. From the bioreactor, the treated groundwater was pumped into a settling chamber from which a portion of the settled sludge was recycled into the

# Index


Activated carbon. *See* Carbon adsorption
Adsorption, 35, 36, 62
  on carbon. *See* Carbon adsorption
Air pollution, 60
Air stripping, 47-61, 79, 146-151
  economics, 52, 151
  high temperature, 49
  in biological reactors, 99-100
Arsenic, 110, 112, 115

Bacteria, 20, 36, 85-87, 97, 101-103, 169-170, 176
  anaerobic, 102
  in air strippers, 61
Bag reactor, 98-99
Bedrock, 4, 5, 35
Biodegradable, 95, 96
Biological treatment, 10, 14, 18-22, 79, 84-109, 145, 161-184
  activated sludge, 18-22, 28, 30-33, 89-93, 107, 108
  economics, 163-164, 183-184
  fixed film, 22, 87, 92, 93, 107, 108
  in-situ, *See* In-situ
  lagoon, 87-89, 107, 108

Carbon adsorption, 11, 25-27, 38, 61-83, 99, 147, 151-161, 163
  breakthrough curve, 68-71
  economics, 74, 75-78, 147, 156, 157, 159, 164
  for air, 60, 79-80
  isotherms, 63-66, 155-156
  mass transfer zone, 70-73
  pressure drop, 72-73
  regeneration, 74-75
  upflow, 72-73

Chlorine, 81
Chlorinated hydrocarbon, 9, 39, 40, 45, 46, 63, 85, 102
  TCE, 9, 26, 65-67, 82
Clarifier, 22, 119-121
  flocculation, 23, 121-122
  lamella, 122-125
  solids contact, 23, 24, 122-123
  thickener, 24, 25
  with bacteria, 20, 89
Cometabolite, 97-98
Conductivity probe, 44-46

Degradability, 38
Density, 7, 8, 11, 37, 38
Distillation, 131-132
Drinking water, 16, 26-27, 61-62, 75, 111, 146-161
Dual media filter. *See* Sand filter

Electrodialysis, 131

Filtration, 78, 124-129
  in carbon bed, 68, 72
Flocculation, 117-119

Gasoline, 8, 36, 38-40, 44, 133-146
Groundwater movement, 2, 3, 5, 13, 41, 46, 142, 146
  hydraulic head, 17

Heavy metals, 86, 87, 100, 101, 110-130
Henry's law constant, 50-52
Hexavalent chrome, 110, 112, 115, 117, 130
Hydrogen peroxide, 81-82

Incineration, 61, 80-81, 85

In-situ treatment, 8, 13–15, 17, 37, 84, 89, 95, 100–109, 173–184
Ion exchange, 24, 25, 129–130

Land farming, 37, 103, 106–107
Landfill, 7, 37
  leachate, 7, 10, 37
Life-cycle
  concentration, 11, 18, 22, 23, 164
  design, 16–34, 165–168
Lime precipitation, 22–25, 110–119

Mist eliminators, 54–55

Nitrates, 111, 130

Oil/water separator, 42–45
Operations
  chemicals, 30–33
  equipment, 30–33
  personnel, 30–33
  power, 30–33
Ozone, 81–82

Packed towers. *See* Air stripping
Packing, 55–58
Petroleum products, 38, 39, 44
Plume, 3–8, 10, 13, 17–18, 26, 103–105
Portable equipment, 29, 120–121, 123–124

Reverse osmosis, 130–131
Ring compounds, 39–40

Sand filter, 24, 125–129
Skimmers, 45
Sludge age, 19–22, 90–91
Slurry wall, 4, 5, 37, 100
Solubility
  metal hydroxides, 113–117
  metal sulfides, 115–117
  organic compounds, 7–9, 11, 36–39
Specific gravity
  bacteria, 20
  organic, 9, 39–40
Spray basins, 47–48
Stripping, 11, 25–27, 38, *See also* Air stripping

Transformation, 7, 9, 10–11, 36
Treatability study, 95–99, 175–176

Ultrafiltration, 80
Unsaturated zone, 2, 5–6, 8, 13, 17, 36, 101–110
UV ozone, 82

Wastewater treatment
  designs, 1, 2, 16, 27
  domestic, 2, 10, 12, 84, 89
  industrial, 2, 10, 12, 84, 89

Zone of influence, 3–4, 14, 104–105